Total Maximum Daily Loads (TMDLs) and Drinking Water Utilities

Awwa Research Foundation

About the Awwa Research Foundation

The Awwa Research Foundation (AwwaRF) is a member-supported, international, nonprofit organization that sponsors research to enable water utilities, public health agencies, and other professionals to provide safe and affordable drinking water to consumers.

The Foundation's mission is to advance the science of water to improve the quality of life. To achieve this mission, the Foundation sponsors studies on all aspects of drinking water, including supply and resources, treatment, monitoring and analysis, distribution, management, and health effects. Funding for research is provided primarily by subscription payments from approximately 1,000 utilities, consulting firms, and manufacturers in North America and abroad. Additional funding comes from collaborative partnerships with other national and international organizations, allowing for resources to be leveraged, expertise to be shared, and broad-based knowledge to be developed and disseminated. Government funding serves as a third source of research dollars.

From its headquarters in Denver, Colorado, the Foundation's staff directs and supports the efforts of more than 800 volunteers who serve on the board of trustees and various committees. These volunteers represent many facets of the water industry, and contribute their expertise to select and monitor research studies that benefit the entire drinking water community.

The results of research are disseminated through a number of channels, including reports, the Web site, conferences, and periodicals.

For subscribers, the Foundation serves as a cooperative program in which water suppliers unite to pool their resources. By applying Foundation research findings, these water suppliers can save substantial costs and stay on the leading edge of drinking water science and technology. Since its inception, AwwaRF has supplied the water community with more than $300 million in applied research.

More information about the Foundation and how to become a subscriber is available on the Web at **www.awwarf.org**.

Total Maximum Daily Loads (TMDLs) and Drinking Water Utilities

Prepared by:
Jeffrey S. Rosen and **Jose A.H. Sobrinho**
Perot Systems Government Services (PSGS)
Mill Wharf Plaza, Suite 208, Scituate, MA 02066
and
Paul L. Freedman and **Wendy M. Larson**
Limno-Tech Inc. (LTI)
501 Avis Drive, Ann Arbor, MI 48108

Jointly sponsored by:
Awwa Research Foundation
6666 West Quincy Avenue, Denver, CO 80235-3098
and
Water Environment Research Foundation
Alexandria, VA 22314

Published by:
Awwa Research Foundation
Advancing the Science of Water®

DISCLAIMER

This study was funded by the Awwa Research Foundation (AwwaRF) and the Water Environment Research Foundation (WERF). AwwaRF and WERF assume no responsibility for the content of the research study reported in this publication or for the opinions or statements of fact expressed in the report. The mention of trade names for commercial products does not represent or imply the approval or endorsement of AwwaRF or WERF. This report is presented solely for informational purposes.

CONTENTS

LIST OF TABLES

FOREWORD

The Awwa Research Foundation (AwwaRF) is a nonprofit corporation dedicated to the implementation of a research effort to help utilities respond to regulatory requirements and traditional high-priority concerns of the industry. The research agenda is developed through a process of consultation with subscribers and drinking water professionals. Under the umbrella of a Strategic Research Plan, the Research Advisory Council prioritizes the suggested projects based upon current and future needs, applicability, and past work, and forwards the recommendations to the Board of Trustees for final selection. The foundation also sponsors research projects through the unsolicited proposal process; the Collaborative Research, Research Applications, and Tailored Collaboration programs; and various joint research efforts with organizations such as the United States Environmental Protection Agency (US EPA), the US Bureau of Reclamation, and the Association of California Water Agencies.

This publication is a result of one of these sponsored studies, and it is hoped that its findings will be applied in communities throughout the world. The following report serves not only as a means of communicating the results of the water industry's centralized research program but also as a tool to enlist the further support of the nonmember utilities and individuals.

The Foundation's staff and large cadre of volunteers, who willingly contribute their time and expertise, manage projects closely from their inception to the final report. The foundation serves a planning and management function and awards contracts to other institutions such as water utilities, universities, and engineering firms. The funding for this research effort comes primarily from the Subscription Program, through which water utilities subscribe to the research program and make an annual payment proportionate to the volume of water they deliver. Consultants and manufacturers subscribe based on their annual billings. The program offers a cost-effective and fair method for funding research in the public interest.

The Foundation's research agenda addresses a broad spectrum of water supply issues: resources, treatment and operations, distribution and storage, water quality and analysis, toxicology, economics, and management. The ultimate purpose of the coordinated effort is to assist water suppliers to provide the highest possible quality of water economically and reliably. The true benefits are realized when the results are implemented at the utility level. The Foundation's trustees are pleased to offer this publication as a contribution toward that end.

The drinking water community has recently expressed the need for a systematic approach to identify and prioritize their role in the TMDL process. There are many reasons why TMDLs may be of interest to drinking water utilities. This project is a starting point to define the role of utilities in the TMDL process and lead to better communication between regulatory agencies and the drinking water communities.

Walter J. Bishop
Chair, Board of Trustees
Awwa Research Foundation

James F. Manwaring, P.E.
Executive Director
Awwa Research Foundation

ACKNOWLEDGMENTS

The authors wish to thank Awwa Research Foundation (AwwaRF) and Water Environment Research Foundation (WERF) for funding this project. The authors thank Jennifer Warner, project manager for the AwwaRF and the following members of the Project Advisory Committee for their guidance and suggestions: Patricia Haddon, Water Environment Research Foundation, Alexandria, Virginia; Donald Catanzaro, FTN Associates, Fayetteville, Arkansas; and Kimberlee Kane, New York City Department of Environment Protection, Valhalla, New York. The investigators also wish to thank all participants in the workshop especially David Katz, Philadelphia Water Department, Richard Gullick, American Water, and Donald Brady, US EPA, Watershed Branch.

EXECUTIVE SUMMARY

BACKGROUND

Total Maximum Daily Loads (TMDLs) are required under Section 303(d) of the Clean Water Act (CWA) for water bodies that do not attain water quality standards after technology-based pollution control requirements are applied. While this provision has existed in the CWA since 1972, it was brought to the forefront through a series of citizen suits in the 1980's and 1990's which required states and United States Environmental Protection Agency (US EPA) to list water bodies not attaining water quality standards (so-called 303(d)-listed water bodies), and develop maximum loadings as necessary to meet water quality objectives. In many cases, aggressive schedules for development of TMDLs for those water bodies were also required.

The Total Maximum Daily Load program focuses on restoring water quality primarily for recreational, aquatic life and public health uses, but addressing water quality concerns is often neglected. This project explores the benefits of drinking water utilities' involvement in the TMDL process. By participating, utilities can leverage the TMDL outcome to reduce costs for treatment, improve water quality, and reduce health risks. This participation will also improve the scope and effectiveness of the TMDL program in meeting all of its wider objectives.

RESEARCH OBJECTIVES

To respond to these needs, the Awwa Research Foundation (AwwaRF) and the Water Environment Research Foundation (WERF) sponsored an experts workshop. They brought together regulators, utilities, and other stakeholders to explore the links between the "clean water" and "safe water" communities represented by CWA and Safe Drinking Water Act (SDWA). The objectives of the workshop were to:

- Identify gaps in current knowledge of the TMDL program
- Identify benefits and impacts to utilities of the TMDL program
- Open communication between regulator agencies and utilities
- Identify future TMDL information and research needs.

APPROACH

To meet the project objectives, a two-day workshop was convened and attended by experts from state and federal regulatory agencies, water and wastewater industries, regional authorities, and nonprofit and non-government agencies. The workshop was held at the River Club in Scituate, MA on December 11-12, 2003.

In preparation for the workshop, a brief survey was designed and standardized to solicit input from workshop participants regarding their familiarity with TMDLs and any specific issues or concerns they might have related to the TMDL process. In addition, a notebook with background materials was sent out to all participants to prepare them for the workshop.

Serving as both management and outreach for this project, an interactive website was designed and developed to store educational materials on TMDLs, pre-workshop materials and project deliverables. The website can be accessed at http://www.drinkingh2o.com/2944. The website will be maintained online and updated periodically for two years (year end of 2005).

PRIORITIZATION OF ISSUES

A major objective of this project was to define the issues related to the TMDL rules and processes from the perspective of a water utility. These issues represent major considerations or barriers that water utilities face when getting involved in the TMDL process. These are challenges that the utilities and the other stakeholders need to be aware of and managed in the TMDL process. The issues were consolidated into six (6) overarching topics including data and analysis, water quality standards versus maximum contaminant levels, implementation of controls, interagency relationships, public outreach, and watershed comprehensive planning.

BENEFITS AND COSTS OF DRINKING WATER UTILITY PARTICIPATION IN THE TMDL PROCESS

The workshop participants recognized and strongly endorsed the importance of drinking water utility participation in the TMDL process and felt that any negative consequences would occur with or without utility participant. The benefits identified were reduced public health risks; reduced treatment requirement and costs; appropriate control measures; good public relations; and other benefits. Workshop participants also identified many potential costs and risks. Costs identified were primarily related to a commitment of time and resources to participation in the TMDL. Risks related to uncertain outcomes of the TMDL were also discussed.

FUTURE RESEARCH

This project identified future research studies that will benefit drinking water utilities. These needs will be valuable for all utilities large and small, regardless of their prior level of sophistication and involvement in regulatory and water quality management issues. The identified needs were grouped in six major areas including TMDL analysis, water quality standards and the TMDL process, implementation of controls, program integration, watershed management, and educating drinking water utilities. Some of the recommended future studies included assessment of state and federal programs integration; interface of CWA and SDWA water quality standards; identification of TMDL data needs for drinking water utilities; compilation of information and questions to be asked on modeling, data, allocation and BMPs; and modeling fate and transport of emerging pollutants.

UTILITY APPLICATION

The results of this research project have numerous practical benefits to the drinking water community. This study presents important information about the TMDL regulations and the relationships of TMDLs to SDWA regulations. The material developed can help guide drinking water utilities into both the liabilities presented by the TMDL requirements and the way TMDLs can be used as part of a source water protection program. This research also identified benefits and risks to water utilities from the establishment of TMDLs and future TMDL research needs that will support the drinking water utilities.

CHAPTER 1
INTRODUCTION

The Total Maximum Daily Load (TMDL) program focuses on restoring water quality primarily for recreational, aquatic life and public health uses, but addressing drinking water quality concerns is often neglected. This project explores the benefits of drinking water utilities involvement in the TMDL process. By participating, utilities can leverage the TMDL outcome to reduce costs for treatment, improve water quality, and reduce health risks. This participation will also improve the scope and effectiveness of the TMDL program in meeting all of its wider objectives.

To respond to these needs, the Awwa Research Foundation (AwwaRF) and the Water Environment Research Foundation (WERF) sponsored an experts workshop. They brought together regulators, utilities, and other stakeholders to explore the links between the "clean water" and "safe water" communities represented by the Clean Water Act (CWA) and Safe Drinking Water Act (SDWA). The objectives of the workshop were to:

- Identify gaps in current knowledge of the TMDL program
- Identify benefits and impacts to utilities of the TMDL program
- Open communication between regulatory agencies and drinking water utilities
- Identify future TMDL information and research needs.

The following sections provide background on the TMDL program and the relevance to the drinking water industry.

BACKGROUND

Total Maximum Daily Loads are required under Section 303(d) of the Clean Water Act (CWA) for water bodies that do not attain water quality standards after technology-based pollution control requirements are applied. While this provision has existed in the CWA since 1972, it was brought to the forefront through a series of citizen suits in the 1980's and 1990's which required states and the United States Environmental Protection Agency (US EPA) to list water bodies not attaining water quality standards (so-called 303(d)-listed water bodies), and develop maximum loadings as necessary to meet water quality objectives. In many cases, aggressive schedules for development of TMDLs for those water bodies were also required.

Existing Situation

Because of recent court rulings, TMDLs have been moved to the forefront of water pollution control activities in most states. Several court rulings that have been handed down have required aggressive schedules for TMDL development. TMDL activities are now being conducted in virtually all US states and territories. According to 1998 statistics, 21,845 water bodies have been listed as impaired on 303(d) lists, and 41,318 TMDLs are scheduled to be conducted over the next ten to fifteen years (US EPA 2000).

The TMDL process requires several component activities, consisting of:

- Listing
- TMDL Development
- Implementation
- Monitoring

TMDLs begin with what is called the "Listing" process. In this process, States review available water quality data to determine which water bodies are impaired and should be placed on the 303(d) list. The TMDL development process is to be conducted for each water body on the 303(d) list. During the TMDL development, the maximum allowable pollutant loads that will provide attainment of relevant water quality goals are defined, as well as individual allocations for each of the sources that contribute to the total. The implementation step consists of conducting the activities required to reduce loads down to their allocated level, typically consisting of National Pollutant Discharge Elimination System (NPDES) permits for point sources and Best Management Practices (BMPs) for nonpoint sources (NPS). Following implementation, monitoring is conducted to determine whether implementation of TMDL control has resulted in compliance with water quality standards.

Activities That Have Shaped the TMDL Process

Beyond their official regulatory functions, US EPA has initiated numerous activities that support progress on TMDLs. These efforts have been ongoing in spite of the budgetary, legal, and time restraints inherent to the TMDL program. Many of these activities have produced or are in the process of producing published documents, and the highlights relevant to this project are summarized below.

US EPA has issued guidance and protocols for the TMDL process, including the following documents:

- Guidance for Water Quality-Based Decisions: The TMDL Process (US EPA 1991)
- Draft Guidance for Water Quality-Based Decisions: The TMDL Process, Second Edition (US EPA 1999a)
- Protocol for Developing Nutrient TMDLs (US EPA 1999b)
- Protocol for Developing Sediment TMDLs (US EPA 1999c)
- Protocol for Developing Pathogen TMDLs (US EPA 2001)
- 2002 Integrated Water Quality Monitoring and Assessment Report Guidance (US EPA 2002a)

TMDL Federal Advisory Committee Act

In 1996, US EPA convened a committee under the Federal Advisory Committee Act (FACA) to develop recommendations for possible changes to the TMDL program. The FACA recommendations were published in July 1998 (US EPA 1998) and were used as US EPA deliberated about its proposed TMDL rules that were issued in 1999. The report provides detailed recommendations related to identifying and listing impaired waters; the pace and scheduling of TMDL development; the TMDL development process; implementation; and allocations. Also discussed are special challenges, stakeholder involvement, and tribal participation, program/agency cooperation, and federal/state/tribal capacity.

National Research Council Review

US EPA contracted with the National Research Council (NRC) to review how scientific data and information should be used in the TMDL program. The authors provide recommendations related to TMDL program goals and the TMDL process, and suggest changes in the data used and analytical methods employed in TMDLs. The results of this study were published in July 2001 by NRC of the National Academy of Sciences (NRC/NAS 2001). Among the issues discussed in the report are the need for better water quality standards and use attainability analyses (UAAs), improved data, better modeling, better analysis of uncertainty and margin of safety (MOS), better implementation, and the use of adaptive management.

The Consolidated Assessment and Listing Methodology Process

The Consolidated Assessment and Listing Methodology (CALM) was developed to help improve state water quality monitoring and assessment programs. More specifically, the methodology, which was developed with state and other stakeholder input, was intended to:

- Reflect the best thinking of state and US EPA water quality program experts on appropriate approaches to assessment and listing under the Clean Water Act
- Provide information to states to assist in preparation of state methodologies for assessment and listing
- Summarize key technical issues and provide information on available technical information and guidance, as well as to acknowledge important gaps or limitations in current scientific knowledge/methods
- Encourage a greater degree of appropriate consistency among states, while still allowing for flexibility to meet state-specific needs
- Foster collection and use of high quality data and the best available scientific methods in assessment and listing programs

The CALM report was issued in July 2002 (US EPA 2002b). It outlines an iterative process for monitoring and assessment programs.

US EPA Office of Water Integration Project

Under the leadership of four Office of Water Division directors, US EPA recently undertook an effort to identify opportunities to better integrate national water program activities. As a first step, the analysis identified major barriers to program integration as well as opportunities to strengthen programmatic linkages. Several of the critical program linkages already identified in the project's first phase are directly related to the TMDL program. Key linkages identified thus far include "aligning water programs to reinforce an integrated watershed approach, developing monitoring programs that support management decisions, and aligning water programs to implement source controls more effectively". Next, the project assessed what institutional changes (regulatory, policy, programmatic) would be needed to support the recommended integration approaches, and determined how performance goals and measures could be modified to track the efficiency and effectiveness of newly integrated processes and activities. US EPA will be exploring opportunities to make structural changes to the national program to strengthen these critical linkages.

US EPA Twenty Needs Report

The Twenty Needs report (US EPA 2002c) focuses on science needs identified by the NRC, states and tribes, US EPA National and Regional TMDL programs, and others. All twenty needs are considered by US EPA as high priority needs. US EPA regional coordinators and staff were asked to pick their top five of the twenty needs. The selected top five needs included improving information on BMPs; improving modeling tools; providing research and development technical support; making monitoring more program-relevant; and improving support for protecting unimpaired waters from degradation.

Water Environment Research Foundation TMDL Review

The Water Environment Research Foundation (WERF) recently sponsored three projects designed to evaluate the existing TMDL program and provide recommendations for improvement. These projects were focused on: 1) The TMDL Listing Process (Thornton and Freedman 2003), 2) Narrative Criteria and the TMDL Process (Freedman et al. 2003a), and 3) The TMDL Development Process (Freedman et al. 2003b). The WERF Listing Process provided explicit recommendations regarding both how to place sites on the 303(d) list, as well as how to determine how to remove sites from the list once evidence shows that this site is no longer impaired. The Narrative Criteria research project provided guiding principles related to: 1) defining impairment in the absence of numeric standards, 2) determining the sources of narrative criteria impairment, and 3) defining target levels. Finally, the TMDL Development study included research related to numerous aspects of the TMDL program, including:

- Estimating background pollutant loads
- Quantifying nonpoint source loads
- Incorporating urban wet weather sources in a TMDL
- Selecting critical conditions
- Guiding principles for modeling in the TMDL process
- Estimating the TMDL margin of safety
- Approaches to the allocation of loads in TMDLs
- Linking BMPs to load reductions
- Implementation planning, tracking, and adaptation
- Adaptive watershed management in the TMDL process

TMDL Rulemaking

Part 130 of Title 40 of the Code of Federal Regulations, section 130.7, contains the regulations currently governing the TMDL program, which were issued in 1992. The "FACA report" (US EPA 1998) served as the basis for many of the provisions in US EPA's first attempt at detailed TMDL rules that were proposed in August 1999 and finalized in July 2000. Subsequent legislation prohibited US EPA from spending funds in fiscal year 2000 or 2001 to implement the new rule. This rule was formally withdrawn by US EPA in March of 2003 as "unworkable." The US EPA has recently introduced a draft Watershed Rule that is designed to redefine the TMDL rules, however a final and revised TMDL rule is expected sometime in the near future.

Relevance to Drinking Water Utilities

The drinking water community has recently expressed the need for a systematic approach to identify and prioritize their role in the TMDL process. There are many reasons why TMDLs may be of interest to drinking water utilities. Maintaining improved surface water quality will likely result in reduced costs for treatment and a reduced risk to drinking water customers. Improving surface water quality and subsequently the load of the disinfection by-products (DBPs) on treatment processes may reduce the risk of DBPs. This could potentially decrease the amount of chemicals necessary for several processes (coagulation, backwash frequency, etc.) ultimately leading to cost reductions for utilities.

The TMDL process and the related US EPA interest in water quality trading can have unusual effects on drinking water supplies. For example, proposed sediment TMDLs in California's Imperial Valley may result in reduced use of irrigation water from the Colorado River that may then lead to increased drinking water allocations. Novel jurisdictional issues are raised when water transfers across watershed and state lines – as with Colorado River waters – intersect with impairments from selenium or perchlorate. Water utilities need to know the appropriate level of stakeholder participation in TMDLs, even when the sources are not nearby, but lie in distant states.

In addition, the development of a more formal decision-making process between regulators and drinking water utilities would provide the drinking water community with options to become involved in the process and interact with regulators to establish future research priorities. This process could be used by the US EPA to identify, select, and prioritize the role of utilities in the TMDL process. This project is a starting point for this approach and will hopefully lead to additional avenues of communication between regulatory agencies and the drinking water communities.

A recent paper (Schechter, Freedman, and Dilks 2000) discussed the overlap between the TMDL program and Source Water Assessment Plans (SWAPs) required under the Safe Drinking Water Act (SDWA), and the rationale for combining these activities. Compelling reasons to performing SWAPs and TMDLs jointly are enumerated below:

1. *Waterbodies are multiple use:* Most of the freshwaters listed for TMDLs also serve as drinking water supply for some public water systems. While SWAPs do also include groundwaters, which are generally not covered under TMDLs, most states have previously completed wellhead protection plans, and thus will most likely be concentrating their SWAP efforts on surface waters.

2. *Pollutant sources are similar:* The top five causes of impairment for a TMDL are very similar to what would be expected as the top five drinking water concerns for a SWAP. The pollutants of concern are similar, and the sources of pollutants are also similar.

3. *Technical requirements are similar:* The technical requirements for SWAPs and TMDLs are extremely similar and performing them together could take advantage of the similarities. Water quality monitoring, mapping and Geographic Information System (GIS), land use analysis, and water quality modeling issues all arise in both SWAPs and TMDLs. These technical requirements often make up the bulk of the effort in developing the plans.

4. *Both programs lack sufficient funding:* The TMDL program will require over 40,000 TMDLs to be developed over the next 8-15 years. Even assuming a lower-than-current cost of $10,000 per TMDL, this yields a total cost of $400 million. SWAPs will be required for the 170,000 water systems in the U.S. While the costs to develop SWAPs are projected to be lower than for TMDLs, estimates of $100 million + are still considered reasonable.

5. *Public notice and public participation requirements are similar:* Both programs require some degree of public input and notice. While professionals may wish otherwise, the general public does not often distinguish between ambient water quality and drinking water quality. Since it is inevitable that questions such as "How does this affect my drinking water?" will come up for a TMDL, why not instead perform the assessments together?

6. *US EPA regions are encouraging states to link the assessments:* US EPA regions have been requesting that states consider TMDLs when developing their SWAPs. While few states have performed this linking to date, it is expected that future SWAPs will incorporate TMDL information.

REPORT ORGANIZATION

The objective of this project was to identify gaps in the current knowledge regarding TMDLs and drinking water utilities, identify the benefits and impacts to water utilities from the establishment of TMDLs, to open the lines of communication between regulatory agencies and utilities regarding TMDLs, and to identify research needs. This introduction is supported and further defined and explained in the following five chapters and eight appendices:

- Chapter 2 presents Methods and Materials which details the approach taken to accomplish the four main objectives of the project;

- Chapter 3 prioritizes and discusses issues of concern to drinking water industry for participating in the TMDL process. Six issues were identified as highest priority in the open discussion during the workshop.;
- Chapter 4 provides the purpose of the benefit and cost analysis and also discusses hypothetical TMDL scenarios;
- Chapter 5 discusses the main categories of identified needs and provides research team recommendations;
- Chapter 6 discusses the relevance of this project to the drinking water industry.
- Appendix A presents the workshop agenda;
- Appendix B provides background presentations conducted during the workshop by one project team member and three workshop participants;
- Appendix C presents the charge for discussing identified issues;
- Appendix D provides the charge for discussing the hypothetical case studies;
- Appendix E provides information the project website;
- Appendix F provides a list of case studies discussed during the workshop; and finally
- Appendices G and H provide the PowerPoint presentations on the costs and benefits and the research needs.

CHAPTER 2
METHODS AND MATERIALS

To meet the project objectives, a two-day workshop was convened and attended by experts from state and federal regulatory agencies, water and wastewater industries, regional authorities, and nonprofit and non-government agencies. The workshop was held at the River Club in Scituate, MA on December 11-12, 2003.

PANEL SELECTION

Workshop participants were chosen based not only on their expertise and experience in the TMDL process, but also on their willingness and availability to participate in the workshop. Based on inputs from the AwwaRF project manager and the project advisory committee (PAC) members, as well as recommendations by the project team, a preliminary list of workshop invitees was established. Invitation letters were sent to all invitees included on this list. Included with the invitation letter was the preliminary list of workshop invitees and a pre-workshop survey designed to assess the level of each participant's familiarity with the TMDL program. The section below describes the details of the pre-workshop survey. Table 2.1 lists the 22 invitees that accepted the invitation and participated in the workshop. Table 2.1 also lists other workshop attendees including the AwwaRF Project Manager, three PAC members, and five project team members.

During the panel selection an effort was made to ensure that the workshop participants were geographically distributed across the United States. It was expected that representatives from different regions of the country would bring to the workshop a variety of issues related to the TMDL process. As shown in Table 2.2, all four US regions were well represented having at least two representatives. It was also important during this process to select individuals that represented different backgrounds. To ensure background diversity the project team targeted individuals from three categories including state and federal regulators, drinking water and wastewater utilities, and "others" represented by nonprofits and non-government organizations (NGO), and regional authorities (Table 2.1). As shown in Table 2.2, the workshop was composed of 13 representatives from water and wastewater utilities, six representatives from state and federal regulatory representatives, and three representatives from regional authorities, non-government or non-profit organizations.

PRE-WORKSHOP SURVEY AND BACKGROUND MATERIALS

In preparation for the workshop, a brief survey was designed and standardized to solicit input from workshop participants regarding their familiarity with TMDLs and any specific issues or concerns they might have related to the TMDL process. The survey also asked the participants to identify research needs and case studies that could potentially highlight the significance of the TMDL program to the drinking water industry.

The responses collected from the survey assisted the project team in planning the workshop and identifying case studies and research needs that were used as a starting point for discussions. The information compiled through the survey also provided the project team with valuable information to ensure that the workshop was focused on the participants' needs and interests.

In addition, a notebook with background materials was sent out to all participants to prepare them for the workshop. These materials are also available online at the project website at www.drinkingh2o.com/2944.

WORKSHOP STRUCTURE

The workshop involved a combination of background presentations, open discussions, and breakout groups. The workshop agenda, as shown in Appendix A, describes the structure of the two-day meeting.

Background Presentations

Four background presentations were conducted at the project workshop by one of the project team members and by three workshop participants. These presentations provided not only a broad overview of the TMDL program and research needs but also the US EPA and utility perspectives on the drinking water utility involvement in the TMDL process. The presentations are briefly summarized in the following paragraphs. The complete four Power Point presentations are included in Appendix B of this report.

1. TMDL Basics for Drinking Water Utilities by Paul Freedman, Vice President, Limno-Tech, Inc., President, Ann Arbor, MI (Appendix B1)

The information presented was tailored for the audience and the objectives of the workshop. It included an overview of the history of the TMDL program, a description of the steps in a TMDL, and recent activities and developments related to the TMDL program. Some of the topics addressed were:

- Overview of steps in a TMDL: listing, development, and implementation
- TMDL requirements
- TMDL conditions including causes, sites, and schedule
- CWA: goals, NPDES program
- Water quality standards
- Drinking water issues
- Research needs

2. TMDL Program – Regulator Perspective by Donald Brady, Chief of the US EPA Water Branch (Appendix B2)

The status of the TMDL program and the agency's perspective on the importance of utility participation in the TMDL process were addressed focusing on the following:

- US EPA TMDL rule: status and lawsuits
- TMDL process: players and roles (US EPA offices and states)
- Future of maximum contaminant levels (MCLs) and water quality standards (WQS)
- How and to what extent drinking water utilities should get involved in TMDLs
- Watershed planning

3. TMDL Program – Water and Wastewater Utility Perspective by David Katz (Appendix B3), Philadelphia Water Department and Richard Gullick (Appendix B4), American Water

The two presentations discussed the importance of water and wastewater utilities participation in the TMDL process including experiences, issues and concerns. Major topics discussed were:

- Why should utilities be involved in TMDLs
- Pollutants of most concern and sources
- Ways the TMDL process could be improved to better protect drinking water sources
- Concerns/fears from the utility perspective if involved in the process
- How and what utilities can contribute to the process
- Conflicts between wastewater and drinking water utilities during the process
- Challenges utilities would face during the TMDL process including technical, policy, and regulatory challenges

Open Discussions

During the first and second day of the workshop, two open discussions involving all participants were moderated by members of the project team leading to a prioritized list of issues and research needs. The workshop agenda in Appendix A lists and describes briefly the main objectives of these two open discussions. The list of issues and research needs were based upon the input provided by participants through the pre-workshop survey, and also input provided during the two workshop sessions. Prioritization of issues and research needs are discussed and detailed in Chapters 3 and 5 of this report.

Breakout Groups

Prior to the workshop, the project team divided the participants into two sets of four breakout discussion groups—one set for each workshop day. Each group consisted of six or seven individuals including one to two regulators, three to four representatives from utilities, one representative from regional or non-profit agencies and one PAC member.

The charge to the breakout groups was provided on paper, and explained with time for questions and answers. The charges for discussion of prioritized issues and hypothetical case studies are included in Appendices C and D.

During the breakout sessions on day one, participants were asked to discuss the six highest-priority issues that resulted from the workshop's Identified Issues session (See Chapter 3). Two groups were assigned to discuss three of the identified issues and the other two groups discussed the three other issues. On day twp, each group was asked to assess two of four possible hypothetical case studies and evaluate the costs and benefits for water utilities that become involved in the TMDL process (See Workshop Agenda in Appendix A). The outcomes of these discussions are presented in Chapter 4.

On both days, a participant from each group was asked to serve as the reporter. The project team also had one representative in each group to take detailed notes. At the end of two sessions, the reporter from each group had 10 minutes to present the outcomes of his or her group discussions.

PROJECT WEBSITE

Serving as both management and outreach for this project, an interactive website was designed and developed to store educational materials on TMDLs, pre-workshop materials and project deliverables. The website can be accessed at http://www.drinkingh2o.com/2944. The website will be maintained online and updated periodically for two years (year end of 2005). The main components of the project website are described in Appendix E.

The website enables project team managers, PAC members, workshop participants, and AwwaRF staff and subscribers to access pertinent project data on TMDLs including workshop materials, important reference documents, US EPA policy documents and guidelines, workshop results, and project reports. This website will serve as a central location for AwwaRF subscribers and stakeholders to obtain current, concise, and comprehensive information on TMDLs and how it specifically relates to their issues.

Table 2.1
Workshop Participants List

Name	Organization	Category	City and State
Donald Brady	U.S. EPA Headquarters Watershed Branch, OWOW	Regulator	Washington, D.C.
Thomas Stiles	Kansas Dept. of Health and Environment	Regulator	Topeka, KS
Teresa Gonzales	California Regional Water Quality Control Board	Regulator	Palm Desert, CA
Richard Gullick	American Water	Utility	Voorhees, NJ
Bradley Brogen	Michigan Public Health Institute		Lansing, MI
Leigh Woodruff	US EPA Region 10	Regulator	Boise, ID
Janice Skadsen	Ann Arbor Water Utilities	Utility	Ann Arbor, MI
Cy Jones	Washington Suburban Sanitary Commission	Utility	Laurel, MD
Jason Wen	Utilities Division, City of Downey	Utility	Downey, CA
Woody Frossard	Tarrant Regional Water District	Utility	Fort Worth, TX
John Troutt	Tampa Bay Water	Utility	Tampa Bay, FL
David Katz	Philadelphia Water Department	Utility	Philadelphia, PA
Rosemary Menard	Portland Water Bureau	Utility	Portland, OR
Marie Light	Tucson Water Department	Utility	Tucson, AZ
Larry Lloyd	Beaver Water District	Utility	Lowell, AR
Edward Graham	Metropolitan Council of Governments	Other	Washington D.C.
Walton Poole	American Clean Water Foundation	Other	Boise, ID
David Williams	East Bay Municipal Utility District	Utility	Oakland, CA
Reed Oberndorfer	Central Utah Water Conservancy District	Utility	Orem, UT
Michelle Farmer	Orange Count Sanitation District	Utility	Fountain Valley, CA
James Greenfield	US EPA Region 4	Regulator	Georgia, GA
David Terry	MA Department of Environmental Protection	Regulator	Boston, MA
Donald Catanzaro	FTN Associates	Project Advisory Committee Member	Fayetteville, AR

(continued)

Table 2.1 (continued)

Name	Organization	Category	City and State
Patricia Haddon	Water Environment Research Foundation	Project Advisory Committee Member	Alexandria, VA
Kimberlee Kane	NYC Department of Environmental Protection	Project Advisory Committee Member	Valhalla, NY
Jennifer Warner	Awwa Research Foundation	AwwaRF Manager	Denver, CO
Paul Freedman	Limno-Tech Inc.	Project Team	Ann Arbor, MI
Wendy Larson	Limno-Tech Inc.	Project Team	Ann Arbor, MI
Jeff Rosen	Perot Systems Government Services	Project Team	Scituate, MA
Jose Sobrinho	Perot Systems Government Services	Project Team	Scituate, MA
Jennifer Ghiloni	Perot Systems Government Services	Project Team	Scituate, MA

Table 2.2
Geographical Distribution of Workshop Participants

Category	US region				Total
	Northeastern	Southeastern	Northwestern	Southwestern	
Utility	4	3	1	5	13
Regulator	1	2	1	2	6
Other	1	1	1	0	3
Total	6	6	3	7	22

*Mississippi River was considered as the East-West boundary line

CHAPTER 3
PRIORITIZATION AND DISCUSSION OF ISSUES

A major objective of this project was to define the issues related to the TMDL rules and processes from the perspective of a water utility. If a water body that is being used as a drinking water source becomes the focus of a TMDL process then the drinking water utilities using this water are likely a major stakeholder. The water utilities can end up viewing the TMDLs from both the source protection perspective as well as contributor to the impaired water body designation. Regardless of the perspective that the water utility has in any specific TMDL process, they need to have a seat at the negotiation table and they also can have a major influence on the process given their expertise, data access and responsibility to their customers.

In order for the water utilities to participate in the TMDL process effectively they need to understand the process, the rules and the options as they engage in the process. The workshop included a significant discussion of the issues that the water utilities needed to face and consider in order to be effective stakeholders in the TMDL process. These issues represent major considerations or barriers that water utilities face when getting involved in the TMDL process. These are challenges that the utilities and the other stakeholders need to be aware of and managed in the TMDL process. The issues were consolidated into six overarching topics with a number of more detailed subjects.

The major topics are listed below, and are explained and elaborated on in the following paragraphs.

1. Data and analysis
2. Water quality standards versus maximum contaminant levels
3. Implementation of controls
4. Interagency relationships
5. Public outreach
6. Watershed comprehensive planning

METHODS OF IDENTIFYING THE PRIORITY ISSUES

The approach taken to identify the issues was based on educating the workshop participants in the TMDL process while contrasting and relating the TMDL process to other regulations and management strategies with which the participants were likely to be more familiar. Educational background materials were sent out to all the participants prior to the workshop. These materials were then summarized and discussed on the first day of the workshop.

Participants were also asked to complete a survey meant to summarize their issues and perspectives prior to attending the meeting. The purpose of the survey was to assess the level of expertise in the workshop participants and to identify possible TMDL case studies, issues and research needs.

Once at the workshop, following the background orientation, the participants were led through a brainstorming activity to capture a wide array of potential issues related to TMDLs from a utility's perspective. Once the potential issues were identified, the workshop group worked on consolidating the issues into a smaller list. Many issues were subsets of others or were different ways of stating the same issue. The issues were combined and the most effective description was applied. Once this step was completed, the group prioritized the six major issues listed above.

Each of the resulting issues was discussed at length in workshop sessions in the afternoon of the first day. Each issue was discussed by two smaller working groups. The working groups focused on the issues by addressing five major points. The five points for discussing the prioritized issues are listed below and in Appendix C. The workshop participants were asked to identify specific barriers that exist for the water utilities to address the issues identified.

- Relevance to a water utility – how could the issue be relevant to a water utility that is faced with participation in a TMDL process.
- Process gaps – identify relevant tools or processes that are barriers or challenges to addressing the issues.
- Knowledge gaps – knowledge, expertise, or fundamental understanding of processes, science, modeling that are a barrier or a challenge to water utilities in participating in TMDLs
- Research needs – what research should be recommended that would help the utilities and other stakeholders effectively participate in TMDLs
- Relevant case studies related to the issues that highlight the issues and both the effects of the issue and possible approaches to address the issues.

The objective of the discussion of the issues was to highlight challenges to drinking water utilities in participating in TMDLs. The structuring of the discussions resulted in clear understanding of the challenges, identification of solutions and of research that could be done to help the utilities better deal with and use the TMDL process.

DISCUSSING THE PRIORITY ISSUES

A clear understanding of TMDL regulations and processes is critical to water utilities. Utilities need to understand when a TMDL is appropriate and the options for developing and implementing and managing a TMDL. Insights into the different roles the utilities could play and the pitfalls that utilities could stumble into are major issues that they might face when becoming involved in the TMDL process. Few utilities have the expertise to design and implement a comprehensive modeling plan that will allow them to assess the issues, plan for a mitigation plan and then monitor the mitigation plan to see if it is achieving its intended effects. A summary of the regulations, options and strategies is needed. The preliminary list of issues included approximately 40 issues that were consolidated into 6 major focus areas. Each of these areas includes a number of different considerations. The issues and sub-topics are discussed below.

Data and Analysis

Any TMDL process requires the collection, organization, documentation, validation, analysis and interpretation of data. How the data are collected and managed are critical to the applicability of the data for all applications related to the TMDL process. The processes include the characterization of the system, establishing a framework for the TMDL discussion, and monitoring the effects of the TMDL implementation. The use of data is also central in the development of reasonable models for evaluating options and predicting results and challenges of any approach.

There are significant issues and challenges related to both the data collection and use of the data for a TMDL. It is necessary to determine if the data that are collected are of use to the TMDL process and what significant data gaps exist. Often times these data are not appropriate for TMDL development or the sampling locations are not appropriate for TMDL evaluation and development. The majority of the data collected that are related to drinking water regulations are measured in either source waters or finished waters. Many of these sampling locations are prescribed by legal requirements. While some of the source water data might be relevant to TMDLs little of the finished water data would be relevant. The issues with the source water data are that they are usually collected at or near the intake which limits it use from the perspective of characterizing or monitoring a water body. In addition, to be useful in TMDLs these monitoring data need to be accompanied by available information about the contaminants of concern. The information required includes sources, sinks and the hydrology of the water body and/or the watershed.

In order for these data be useful to the TMDL process, a significant amount of metadata are required (metadata is information that describes the data). It is important to understand:

- Why the data were collected
- Frequency of data collection
- How many stations were sampled
- How were the samples processed
- Methods applied to analyze the samples
- What kind of quality assurance and control was applied to the samples and the data
- How were the data managed, and
- What if any data aggregation has occurred.

Prior to being considered as a resource in a TMDL process the methods and tools for capturing, analyzing, aggregating and organizing data all need to be carefully considered. Since a TMDL requires an estimate of the inputs that overwhelm the capacity of a system some model or method of estimation is necessary to connect the conditions and inputs to the system with causes of impairment for a designated use. The available data then becomes an input to the appropriate model to validate the model for the system in question. If the model appears to be valid for the system then it can be used to evaluate if the interactions of inputs and dynamics of the system are responsible for the failure to meet the designated use. The model can then also be used to evaluate and contrast a variety of TMDL strategies for eliminating the impairment. Therefore there is a close relationship between the data requirements and the models that are used for TMDLs. The right data are needed in the right format for a model that properly describes the system that is to be managed using a TMDL. These relationships between available data and appropriate models are likely knowledge gaps not only for the utilities but for the majority of the stakeholders.

Once a TMDL is developed and agreed upon, the stakeholders need to have a plan for how to collect the data to determine if the TMDL is having the desired effect. Few utilities have the expertise to design and implement a comprehensive plan that will allow them to assess the issues, plan for a mitigation plan and then monitor the mitigation plan to see if it is having its intended effects.

There are water bodies that are on the 303(d) list that may no longer be impaired for designated uses. In a case where a utility may be forced to participate in a complex TMDL process on a water body for which a TMDL is really not needed, a better understanding of how a water body can be de-listed from the 303(d) list would be useful. An understanding of what data and what kind of monitoring is necessary to de-list a water body are not well understood.

Water Quality Standards versus Maximum Contaminant Levels

Historically, drinking water utilities have been mostly concerned with the Safe Drinking Water Act (SDWA) regulations and standards. These are significantly different than the Clean Water Act (CWA) regulations and standards. While the SDWA regulations focus mostly on finished water, the CWA focuses mostly on surface waters. The allowable values established for these two sets of regulations are often completely different because they have very different goals and protect different resources that may be at risk.

Since the SWDA focuses on the requirements for drinking water, they are mostly based on the quality of finished water and on attaining required inactivation for microbiological agents. As a result of these requirements, chemical concentrations in drinking water are usually established as Maximum Contaminant Levels (MCLs) and they are based on human health consideration. The MCLs are an estimate of the amount of a chemical that a person can be exposed to on a daily basis that is not anticipated to cause adverse health effects over a person's lifetime (http://www.epa.gov/safewater/mcl.html#1).

The regulations that are most important in consideration of TMDLs are based on the CWA. These regulations specify water quality standards based on specific designated uses. The water quality standards (WQS) are those concentrations of contaminants above which the water body is no longer appropriate for the designated uses (http://www.epa.gov/watertrain/cwa/cwa4.htm). These can be expressed as concentrations of pollutants, temperature, pH, turbidity units, toxicity units, or other quantitative measures. Drinking water is one possible designated use but water bodies may have multiple designated uses.

In the past, water utilities have always considered the source water concentrations from the perspective of what must be done to bring them under the MCLs. The costs of achieving these goals were considered operating expenses. Some utilities have successfully participated in watershed management to lower the costs of treating water. TMDLs are a mechanism for supporting these watershed management goals. In the case where a water utility is a source of a contaminant they may now be forced to participate and absorb part of the costs to support attainment of uses other than drinking water.

It is important to note that water utilities historically have sampled and monitored with the interest of characterizing and meeting SDWA requirements. Watershed sampling has been limited to required source monitoring (which is minimal) and for determining treatment requirements. The sampling requirements for TMDLs are likely to be significantly different than the sampling that has been done historically. This will require additional sampling and changes in the sampling plans in order to characterize the source water body and to be able to detect differences that will indicate that the TMDL is having an effect. Opportunities for using existing data or redesigning monitoring so that results can be used for both SDWA and CWA monitoring should be explored. The utilities should also initiate discussions with the regulatory agencies to determine how they might leverage existing monitoring to meet multiple obligations.

Drinking water utilities need to consider TMDLs from a number of perspectives:

1. Possible tool to support source water protection and improve source water quality

2. Process that they should involved if any water body that is a potential drinking water source is ruled as impaired

3. Potential new cost both from a participation perspective and from a mitigation perspective

Water utilities should become familiar with the TMDL process. If they use surface water as a source they should determine if their sources have been designated as drinking water sources. For those that have been designated as drinking water sources the utility should determine if they are attaining the use or are they listed as impaired. If they are impaired then the utility should consider leading a TMDL effort to determine the sources and possible mitigation strategies for the impairment. The utilities should not wait for someone else to initiate the TMDL process.

For water bodies that are listed as impaired, water utilities should train technical people in applying and using the CWA to their own benefit. All available data for the water body should be located and organized. Possible sources of impairments and likely stakeholders should be located. Initiation of the process should be facilitated through the appropriate state agency. Water utilities should monitor trends and issues in the regulatory stakeholder process to better understand tools to help them protect their sources and issues that could draw a utility into a TMDL as a possible cause of non-attainment.

Implementation of Controls

Once TMDLs are developed for a water body then the strategy and logistics of the TMDL implementation and monitoring become important to the utility. Decisions may result from the TMDL process that includes mitigation, controls, and other initiatives that theoretically can help a water body improve. Once these recommendations are made they will need to be iteratively implemented, monitored and adjusted. This is not a trivial matter and there are not many organizations that are likely to be part of the stakeholder process capable of this on-going management of the TMDL process. On top of that the implementation and monitoring of the TMDL can be an expensive and technically challenging. If the water utility is at the table as a stakeholder they may be called upon to conduct the sampling, analysis and interpretation of the data or to pay for the implementation. Unless this stage is done properly no one will ever know if the TMDL is having the desired effect. If the water utility is not involved in the implementation then it is important to understand the capabilities of whoever is designated as the implementer.

In cases where the utility is not the party responsible for the impairment it may need to monitor these other sources and then work with the primary agency to ensure that the TMDL is working and that the responsible party is performing its obligations under the agreement in the mitigation plan. Any TMDL developed will face the challenge of resources to perform the mitigations, design and implementation of the monitoring, management of the data, performing appropriate analyses and then communicating the results to the primacy agency.

Utilities should decide what role and what level of involvement they will have in a TMDL related to either their source or receiving water. The implementation of controls can be an opportunity for utilities to lead the process and focus the approach on the desirable effects from the water utility's perspective. It is also an opportunity to develop a cooperative group of stakeholders working together to meet the objectives for mutual benefit. Water utilities will also have an opportunity to work closely with the state agencies in a partnership that could improve communications and reduce some of the difficulty in working with the primacy agencies. Since the stakeholders are likely to look to the utilities for resources to support the TMDL process, the implementation of controls is an opportunity to leverage the process in order to get other stakeholders to share the costs of the TMDL implementation and long term management of the process. If the utility chooses to not take a leadership role they should still take an active role to ensure that the TMDL does not force them into processes or activities that are not in the best interest of the utility or the consumer.

Utilities also need to ensure that if a mitigation plan is put in place the agency capable of enforcing the required actions is identified and the decision points are clearly articulated and agreed to by all parties. If stakeholders are not holding up their end of the process there needs to be an authority that can force the participants to fulfill their obligations. Early discussions should include strategies for removing a water body from the 303(d) list when the TMDL meets its intended goals.

To best leverage the implementation of controls, water utilities should continue to develop watershed management and source water protection plans. Utilities should also make sure that they are in a position to understand the problems and the options available. An understanding of Best Management Practices (BMP) as they relate to managing environmental resources could be useful in the TMDL process. A detailed understanding of how these BMPs are applied and managed and what effects are expected would be useful in the planning and measuring of the TMDLs efficacy. In addition, water utilities should begin gathering innovative approaches and case studies that could be used to ensure that the TMDLs proposed are useful and will have the desired effect.

Interagency Relationships

The emergence of TMDLs as a significant concern for water utilities results in the water utilities coming under regulatory scrutiny from two different perspectives. The first being the drinking water standards and the second being the scrutiny of the watershed and surface water quality perspective. Each state normally has different agencies focusing on these different, but related issues. The need to meet these requirements of multiple agencies regarding the management and protection of source waters can result in conflicts in prioritizing strategies within the utility. Different regulations and goals could be at odds with one another and could incur costs that could be avoided if the utilities could get the regulating agencies to coordinate these two sets of water quality standards. It is critical to keep in mind that these conflicts and the gap between the two sets of water quality standards are legislated and not likely to go away. TMDLs are now getting much more focus than they have in the past and this will likely result in additional requirements for water utilities that will need to be met. As water bodies are evaluated and prioritized for TMDL processing it would be useful for the water utilities to be involved in setting the priorities so that drinking water sources are considered for TMDLs early in the process. This will require interactions between the agencies and an understanding of the priority of the drinking water sources.

Aside from the potential conflicts between agencies responsible for the CWA and the SDWA there are potential conflicts among the different agencies that could have jurisdictions over the TMDL process. Given the watershed nature of the CWA, it is possible for multiple agencies including state, federal and tribes to get involved in TMDLs. The perception of agencies relative to one another regarding who has the authority and what strategies to employ could be a barrier to meet the regulations. As much as possible the utilities would be best served by facilitating cooperation among the regulating agencies.

Water utilities can also work with the different state agencies to determine if mitigation efforts related to source water protection and SDWA consideration can be leveraged to deal with CWA issues. It is possible that State Revolving Funds (SRF) could then be used to deal with some of the costs related to the TMDL process.

Public Outreach

When regulations and processes are related to public drinking water sources, the entire community which may include adjacent watersheds are stakeholders. Discussions that focus on water bodies not meeting designated uses may generate interest from a wide array of constituents and are often topics of intense discussion in the press and other public forums. Since the mitigation of non attainment will often require significant resources, organizing public and political support is important. For all of these reasons, water utilities may wish to make a concerted effort to reach out to the public to communicate the plan for the TMDL development and the long term evaluation and management of the TMDL.

Public outreach will facilitate the identification of the appropriate stakeholders and might help organize a group of advocates both for the required work and for funding. This group of advocates could also help mobilize resources and support for source water protection and on-going source water monitoring. Public outreach and education will also be critical for utilities to communicate best management practices to stakeholders who may feel that they are being adversely impacted. This includes outreach to farmers to help with non point source inputs to source waters.

While public outreach is considered very positive for utilities involved in TMDLs it is important to also recognize some of the challenges that public outreach introduces. Public outreach can be expensive and can take a great deal of time. Usually TMDL development is under severe time constraints due to regulations and resources. Public outreach can lengthen this process. It is also necessary to identify public constituencies that might be interested and what is the best way is to reach contact and interact with them.

Watershed Comprehensive Planning

Faced with divergent and often incompatible interests within watersheds, many organizations, states, regions and agencies have undertaken comprehensive watershed management plans. These plans are efforts to organize all of the issues, regulations, interests and resources and develop a plan for managing the resources, protecting the environment, maintaining reasonable and responsible growth in a framework that all stakeholders can understand and agree to. This is a major effort that encompasses all of the issues addressed above and as well as others. It is a logical framework within which the TMDL process needs to be considered as a component.

One benefit of a comprehensive watershed management plan is that it may help water utilities meet their regulatory requirements for safe drinking water by protecting water sources and decreasing the variability in water quality and reducing contaminants that need to be removed or inactivated during treatment. These plans also help utilities reduce disinfection by-product (DBP) precursors that result in undesirable treatment by-products. As a result, the costs and complexity of water treatment may be considerably reduced.

While the integration of the TMDL process into the comprehensive watershed management plan is a very appealing idea, there are major barriers that need to be considered. First, the Comprehensive Watershed Management approach is not the sole responsibility of any organization, nor is it mandated or funded. Enforcement of the plan needs to be arranged as part of the planning process. In order for it to happen there needs to be a coalescing of stakeholders, a mobilization of resources and then a long term commitment to enforce the plan. Enforcing the plan is sometimes difficult but can be done through zoning, open land programs, state regulations and regulatory incentives. Water utilities sometimes take the lead in developing and organizing comprehensive watershed management plans. TMDLs could help by forcing stakeholders to participate but it also can result in the process getting mired down in bureaucracy and the other issues identified above. In some cases the utility is best just getting the plan in place and then supporting TMDLs separately.

There are some excellent examples of how comprehensive water management plans can be done and the positive effects that they can have. Some of these are listed in relevant case studies below. Most of the references indicate a government agency taking the lead in coordinating and managing the comprehensive watershed monitoring plan. Water utilities can take a leadership role in the process. A good example of this has been the initiative taken by Philadelphia Water Department (see relevant case studies in Appendix F).

Comprehensive watershed planning is a process that has been formalized in a number of locations. For example the state of Washington has a formal Comprehensive Watershed Planning Act that has had some success in initiating plans in multiple watersheds (http://www.mrsc.org/subjects/planning/watershed.aspx). A number of specific efforts are underway in Washington State (http://www.ecy.wa.gov/watershed/index.html). These references are an excellent source of approaches and lessons learned. The case studies also point out some of the problems with this comprehensive watershed planning, specifically, that there is no standardized way to develop a plan. TMDLs are often located in watersheds that are doing comprehensive plans but the comprehensive watershed plan is usually focused on a larger area and resource than what needs to be addressed in a TMDL. The TMDL efforts are usually a part of a larger comprehensive management plan.

This combination can be a great opportunity for a water utility to participate in protecting its water sources while mobilizing resources and helping support long term planning, water quantity planning and source water protection. This is also a good way for a water utility to demonstrate its commitment both on the source side as well as on the discharge side since the comprehensive watershed management planning will address both issues and will treat them as connected, requiring coordinated management.

All the challenges discussed above that are specific to TMDLs are also relevant when comprehensive watershed management is discussed. The access and organization of the required data is a major challenge that is often underestimated. Expanding the TMDL effort into a comprehensive watershed management plan will require a great deal of extra data. A major additional cost is the organization and standardization of Geographic Information Systems (GIS) data coverages. This adds significant cost to the effort. Other challenges include developing and implementing a monitoring plan and bounds within which effects of the plans and mitigation can be evaluated. Combining the comprehensive watershed management plan and TMDLs can make the monitoring more difficult and complex but it also can help with resources for doing the monitoring and data analysis. While this comprehensive and integrated approach makes good sense and will likely lead to a very well designed and implemented TMDL, each case should be looked at independently to see whether or not the combination is likely to be effective. There are likely to be situations in which a simple TMDL is more desirable and feasible.

CASE STUDIES

During the session on prioritization and discussion of issues, several cases studies were provided by workshop participants. These case studies are listed in the previous sections and in Appendix F.

CHAPTER 4
BENEFITS AND COSTS OF DRINKING WATER UTILITY PARTICIPATION IN THE TMDL PROCESS

This chapter provides a *qualitative* exploration of the benefits and costs of drinking water utilities' participation in the TMDL process. A *quantitative* assessment is difficult because values are highly site specific. Furthermore, the outcome of such participation can be uncertain, and case examples are limited. This chapter describes the investigation and workshop elements conducted to address the topic, and highlights the need for more information.

As a first step, the project team reviewed background studies that have explored the relationship between watershed management and drinking water source water protection. The findings of the review were presented to workshop participants in materials provided before the workshop, and in a workshop presentation (Appendix H). Participants were then asked to identify potential benefits and costs through a review of hypothetical TMDL scenarios. This chapter provides a broad picture view of benefits and costs identified through this process, with references and illustrations where possible.

BACKGROUND STUDIES

Some research has been compiled on the relationship of watershed management to drinking water source water protection. TMDLs are in a sense just a regulatory version of a component of watershed management plans. Highlights of some selected studies that explore the relationship between watershed management and source water protection follow. These do not represent a comprehensive review of the literature; rather they provide a sampling of the types of studies that have been conducted. Studies that specifically relate TMDLs to source water protection, with benefits and costs provided, were not located through this literature review. However, this was not unexpected, because drinking water utility involvement in TMDLs remains relatively limited.

A U.S. EPA report, *Protecting Sources of Drinking Water; Selected Case Studies in Watershed Management* (US EPA 1999d) provides case studies of 17 drinking water systems, and discusses lessons learned in four aspects of source water protection: building partnerships, watershed assessment, watershed land use management, and land acquisition. The descriptions of case studies for each aspect are comprehensive but generally qualitative in nature. Nonetheless, this report has direct applicability to TMDL involvement.

One example of a case study in the EPA report highlights the Salem, Oregon Public Works, located on the North Santiam River. Two consecutive summers of high rainfall resulted in persistent turbidity, leading to a disruption of slow sand filtration at the drinking water plant. The city was forced to use alternate water sources, install temporary treatment systems, and curtail use. Through a negotiated agreement with the U.S. Forest Service, joint monitoring programs were established, and many other strong partnerships were established to coordinate watershed protection efforts. Benefits to the utility were not quantified, but they include an increased understanding of the quality of the source water, and identification of the areas in the watershed most in need of management actions. Through formal partnerships, Salem Public Works was able to engage a wide range of stakeholders to cooperate in protecting the health of the source water. The result was improvement and protection of water quality and more consistent treatment operations.

The Salem example is supported by the findings of a recent quantitative study that focused on the multiple benefits of watershed protection through land conservation (Ernst et al. 2003). The article highlights the relationship between forests and clean water, and discusses how drinking water utilities can build partnerships to conserve forested land. The article presents the results of a survey of 27 water utilities, conducted by the Trust for Public Lands and the American Water Works Association's Source Water Protection Committee, which found that the operating cost of water treatment varied depending on the amount of forested or natural lands in the watershed. Specifically, the article noted that for every 10% increase in forest cover (up to about 60% forest/natural cover), treatment and chemical costs decreased by approximately 20%. The analysis allowed prediction of average treatment costs per day (if a supplier treated 22 million gallons per day, the average production of the surveyed utilities), as shown in Table 4.1.

Table 4.1
Predicted Change in Treatment Costs as a Function of Percent Forested or Natural Lands

% of Watershed Forested	Treatment and Chemical Costs per MG	% Change in Costs	Average Treatment Costs per day at 22 MG
10	$115	19	$2,530
20	$93	20	$2,046
30	$73	21	$1,606
40	$58	21	$1,276
50	$46	21	$1,012
60	$37	19	$814

Source: Ernst, Gullick, and Nixon 2004. Protecting the Source: Conserving Forests to Protect Water. *Opflow*, 30(5):1-7. Published with permission from AWWA.

In addition to these predicted estimates of treatment cost savings, the article provides two case studies where drinking water utilities have partnered with land trusts and other parties to implement source protection strategies. The case studies illustrate the importance of land conservation as part of a multi-barrier approach to providing safe drinking water.

Quantitative information on benefits of watershed management is also provided in a report titled *Valuing Benefits from Water Management* (Conservation Ontario 2001). The study was conducted as part of the Canadian Walkerton Inquiry in response to an outbreak of *E. coli* in Walkerton, Ontario. The report provides a summary of the benefits and value of watershed management, based on a literature review of 13 U.S. and Canadian papers. While the study focuses on protecting water supply, the authors acknowledge that there are multiple objectives and benefits resulting from watershed management. They point out that many problems are the result of diverse sources, and solutions rely on land and water resource management that is beyond utility control, so the involvement of multiple stakeholders is essential. The report lists several watershed management actions that may provide benefits, including agricultural measures, wastewater treatment plant upgrades, urban storm water management, modified operation of reservoirs, protection of natural areas, and water conservation. Table 4.2 lists some categories of benefits and estimated dollar values for benefits that may result from these actions.

Table 4.2
Estimated Dollar Values of Benefits of Watershed Management Actions

Category of Benefit	Estimated Value (2001 $U.S.)
Avoidance of losses due to infection by *Giardia*	Individual loss: $2,500/case Total cost of actions to avoid: $900/household Loss of business income due to outbreak: $12,000/establishment
Prevention of tastes and odors by controlling precursors	Annual willingness to pay: $130 per urban household; $170 per rural household (private well)
Impact of poor source water quality on treatment costs for 430 of the 600 largest utilities in the U.S.	1% increase in turbidity increases costs by average 0.07% ($113/million gallons)
Delay nitrate contamination of GW by 20 years by controlling waste from hog farms (Iowa)	Annual willingness to pay: $95 per household

Source: Data from Conservation Ontario 2001.

The Walkerton report also provides several case studies to illustrate the range and magnitude of values related to benefits. One high profile case study describes a watershed strategy to protect New York City (NYC)'s water supply: the Catskill/Delaware Watershed Management for NYC Water Supply System. Key program elements include land acquisition (up to 355,000 acres), disease surveillance for Giardiasis and Cryptosporidiosis, a program to protect reservoirs from high phosphorus loadings, agricultural, forestry and storm water pollution control programs, and financial support to watershed communities. Planned investments through 2007 are $1.5 billion, and proponents claim these measures will eliminate the need for investment of approximately $8 billion in a conventional filtration plant.

Quantitative estimates of benefits are also provided in a study by Dearmont and others titled, *Costs of Water Treatment Due to Diminished Water Quality: A Case Study in Texas* (Dearmont, McCarl, and Tolman 1998). Data for a three-year period for 12 treatment plants in Texas were evaluated. Chemical costs were expressed as a function of raw surface water quality, and sediment was used as an indicator of water quality. A 1% increase in turbidity was predicted to increase chemical costs by 0.25%, or $534 annually for the average plant and $69,826 annually on a statewide basis. When contamination was present, the chemical cost of treatment was found to increase by $95 per million gallons from a base of $75. Consideration of the effects of turbidity and contamination on non-chemical costs would raise these damage estimates.

While these studies do not represent a comprehensive review, they highlight some potential benefits and costs of utility involvement in watershed management activities. Brief summaries were presented to workshop participants to set the context and provide background for discussion by workshop participants.

WORKSHOP APPROACH

Within the two-day workshop, a half-day session was focused on the topic of benefits and costs of involvement in TMDLs. In a brief presentation (see Appendix G), participants were first provided with some background information, including summaries of the literature studies described in the previous section. Participants were then asked to examine and discuss four hypothetical TMDL scenarios involving drinking water utilities, and identify through small group discussions potential benefits and costs of participation in the process. They were asked to put themselves in the role of a utility employee who needs to collect information related to benefits and costs in order to receive permission to get involved in the TMDL. Two of the scenarios involve a utility impacted by impaired source waters (nutrient and pathogen TMDLs) and in the other two scenarios (dissolved oxygen and temperature TMDLs), the utility contributes to the impairment. An overview of the four scenarios follows; complete descriptions are found in Appendix D.

Nutrient TMDL

You are on the regulatory staff at a utility that draws water from a small river. The watershed is large, and 80% of the land is used for agriculture. Your utility has had filtration problems and chronic taste and odor problems in finished water due to algal blooms in the reservoir. Septic systems may also contribute phosphorus to the system. The river was declared impaired due to excessive algae caused by high phosphorus loadings.

Pathogen TMDL

You are on the regulatory staff at a drinking water utility that services a large metropolitan area. Water is drawn from a reservoir on a large river. The watershed is 40% urban and 60% rural. Illicit connections to the stormwater system and sewer breaks are common, and there are many CSOs (combined sewer overflow) and SSOs (sanitary sewer overflow). You have had chronic problems and concerns related to pathogens. The river was determined impaired because bacterial concentration exceeded recreational contact standards.

Dissolved Oxygen TMDL

You are on the regulatory staff at a drinking water utility that services a large and growing urban area. The intake is located on a large shallow river in a largely forested watershed. Development in the watershed has reduced the percentage of forested land, and increased the percent urbanized area to 25%. The wastewater treatment plant for the community is located in a different basin, so there are no return flows to the river. You anticipate increasing your withdrawal rate in the future due to expected population growth.

Hydrologic conditions that affect surface water quality include high variability in streamflow, characterized by extreme low flows in the late summer-early autumn seasons and magnified peak flows due to high impervious land cover in the urbanized areas. Your intake contributes to the summer low flows. There are extreme diurnal swings of dissolved oxygen in the river, including below your withdrawal (gets worse as go downstream), and nuisance growths of periphyton have been observed in low-flow periods. BOD loads from urbanized areas contribute to the problem. Dissolved oxygen concentrations routinely fall below the state water quality standard to protect fish.

Temperature TMDL

You are on the regulatory staff at a drinking water utility for a large metropolitan area that draws its water from impoundments on a large river system. The drinking water dams generally release cooler-than-natural water in early summer months and warmer water during late summer months when the supply of cold water is exhausted. Reduced flow volume below the dams from dam operations also causes high stream temperatures downstream.

The river supports several indigenous cold water fish species including Chinook and Coho salmon and Steelhead trout. These fish need cold waters for optimal health at various times during their life cycle. The State's stream temperature standards are designed to be protective of these needs. Exceedances of the temperature standard have led to the river (including the reservoirs) to be listed for temperature.

Provided below is a summary of the benefits as well as the costs and risks identified by workshop participants.

POTENTIAL BENEFITS OF UTILITY INVOLVEMENT

In general, workshop participants strongly endorsed utility involvement in TMDLs. They generally argued for a proactive stance, taking a leadership role in shaping the outcome of TMDLs to best advantage, rather than taking a reactive stance by only responding to or benefiting from the outcome. The desire on the part of the utility would be to insure that any controls that are imposed as an outcome of the TMDL are maximized to best protect source water. Additionally, if there is a possibility that controls might be imposed on a utility, it is important to be proactive to ensure the proposed controls are necessary, appropriate, feasible, and cost-effective.

The following sections describe the categories of benefits identified through the workshop. Participants were asked to identify benefits through examination of the four diverse TMDL scenarios to evoke discussions from broad viewpoints. While some of the benefits identified through this process were specific only to the particular scenario under discussion, most responses can be universally applied to all TMDLs. Therefore the benefits identified through the breakout sessions are grouped herein by general categories of benefits, rather than by TMDL scenario. The categories are: reduced public health risks; reduced treatment requirements and costs; appropriate control measures; good public relations; and other benefits.

Reduced Public Health Risk

Much discussion centered on the topic of reduced public health risks as a benefit of TMDLs since public health protection is the paramount objective for utilities. The leading public health risks in water supply overlap with some of the top causes of impairment in the TMDL program, therefore it is logical that controls implemented through the TMDL program may benefit utilities by reducing public health risks. TMDLs can reduce risks through higher quality source water, and/or by lowering the risk of episodic contamination. Reduced public health risk may result directly from the TMDL through lower bacteria levels in the source water for example, or indirectly through reduction of associated parameters that are not directly regulated by the TMDL.

Three of the top four causes of impairment in the TMDL program are sediments, bacteria, and nutrients, amounting to 40% of the total required TMDLs nationwide. Reducing bacterial pollutant loads through a TMDL has an obvious benefit to reducing public health risks. However, reducing sediment loads through TMDLs also reduces public health risks. Reducing sediments reduces turbidity, a surrogate for the presence of pathogenic bacteria, *Giardia*, and *Cryptosporidium*. Reducing nutrient loads through TMDLs can reduce risks by reducing natural organic matter (NOM), precursors to disinfection by-products that can be carcinogenic. Hence, reducing nutrients can reduce NOM, which can reduce cancer risks.

In addition to sediment, nutrient, and pathogen controls, the workshop participants discussed the TMDL reduction of other contaminants such as synthetic organic pesticides, metals, mercury, and PCBs (polychlorinated biphenyls) that can also reduce public health risks. These types of contaminants may be difficult to treat at the water plant, and the public water supply will benefit from reduced levels in source water. Contamination by metals, pesticides, and other toxic chemicals amount to approximately 20% of the impaired waters, so these TMDLs can be important to drinking water utilities. Additionally, public health risks may be reduced indirectly by a reduction of contaminants that are not directly regulated through the TMDL. For example, nutrient controls as a result of a nutrient TMDL will reduce natural organic matter and can lower the concentration of disinfectant by-products (DBPs) such as trihalomethanes (THMs) in the water supply. Another indirect example is the case of a reduction of solids runoff from farms as a result of a sediment TMDL, which may reduce pesticides in the source water and/or pathogenic risks.

Reduced Treatment Requirements and Costs

Workshop participants also discussed the very direct potential benefit of a TMDL that could reduce treatment requirements and/or operational costs. Over 25% of the required TMDLs involve sediment and nutrients which contribute to three common operational problems at utilities, taste and odor, high suspended solids, and high turbidity. Controlling watershed sources of sediments and nutrients can reduce utility operational requirements for extra filtration, increased frequency of filter backwashes, and costly chemical dosages. In addition, savings might be gained by eliminating the need for specialized taste and odor controls, such as ozonation or additional filtration requirements. For these financial reasons alone, workshop participants spoke strongly about the importance of utilities advocating their interests in these types of TMDLs.

Appropriate Control Measures

Although most of the participants' discussions focused on better regulation and control of watershed sources of pollution, some TMDLs can result in regulation of drinking water utilities and these situations were also discussed. Control measures may be imposed on a utility as a result of the TMDL process, particularly when a utility is contributing to the impairment (as in the dissolved oxygen and temperature TMDL scenarios). In these cases, utility participation in the TMDL process protects the interests of the utility, and insures that any imposed controls are fair and feasible. Controls that may be imposed include relocating a water source, relocating a discharge, treating a discharge of treatment solids or backwash water, and reducing the flow rate of a water intake. All of these controls can be costly or restrictive to a utility. Therefore, being actively involved in the TMDL process best protects the utility's interests.

Good Public Relations

The benefit of good public relations was an outcome of TMDL participation often touted by workshop participants. Improving finished drinking water through TMDL improvements to the source water can improve public confidence and improve customer satisfaction. Some benefits are obvious, such as in the case of taste and odor problems from algal blooms. There may be savings in terms of the time spent by staff in handling complaints from customers. However, improved water quality may also improve recreation opportunities, resulting in improved public relations with the community, an important asset when discussing rate changes. Also, the stakeholder involvement component of the TMDL process provides a forum to interact with the general public. This provides an opportunity to strengthen relationships, hear concerns, publicize the good work of the utility, and educate the public about drinking water issues of importance. Overall, the workshop participants felt if nothing else is gained from TMDL involvement, the benefit of good public relations warrants participation.

Other Benefits

Participants also frequently spoke of the secondary benefits of TMDL involvement, including more efficient regulatory interactions, improving water quality information, and more effective future planning.

Regulatory interactions on any issue can be costly to utilities. The TMDL may address other utility concerns and potentially allow for multiple issues to be addressed through one set of negotiations and public hearings. This naturally results in more efficient regulatory involvement. For example, the TMDL may also be tied into other environmental issues of concern, such as issues related to the Endangered Species Act (ESA), or water withdrawal restrictions. Addressing all such issues together leads to more effective solutions and a more efficient process.

The TMDL process may also result in monitoring enhancements involving multiple parties, which will expand the amount of data available for utility management decisions. Through participation in the TMDL, utilities may be able to design sampling to capture additional issues that are relevant not only to the TMDL but also for the utility. The monitoring design may include better characterization of pollutant sources as well as posing risks to the water supply. Monitoring data collected cooperatively by multiple parties may also provide more information on source water quality, leading to more informed decisions about locating water intakes or seasonal use patterns as examples.

Finally, future planning for the utility may become integrated into long-term watershed management planning through the TMDL process. For example, a utility that serves a growing community may need to increase their water withdrawal at some time in the future. By becoming involved in the TMDL process, there may be opportunities to affect water conservation, and other measures that may stem from a comprehensive watershed management plan. In some cases, protections and restoration of the ecosystem may result in a delay or avoidance of moving to alternate sources.

POTENTIAL COSTS AND RISKS OF TMDL INVOLVEMENT

The perceived down side of TMDL participation was generally not considered significant to workshop participants who felt that any negative consequences of TMDLs would occur *with or without* utility participation. In general, they felt that these consequences are better mitigated or directed through active participation in the process. However, there are definite costs and risks of involvement. Costs identified through the workshop were mostly related to a commitment of time and resources to participation in the TMDL. Risks related to uncertain outcomes of the TMDL were also discussed.

Costs of Participation

Participation in the TMDL process will require attendance at meetings but might also include the need for information research and data collection analysis. The time commitment, resources, and associated costs for these activities may be significant, as discussed below.

Attendance at Meetings

Participation in meetings and interactions with other stakeholders may be quite time-consuming because of the large number of issues and stakeholders involved. TMDLs may engage many stakeholders, including Federal agencies such as the Corps of Engineers (USACE) and US Environmental Protection Agency (US EPA), state environmental and health agencies, and local entities as well as non-governmental organizations (NGOs) such as the Farm Bureau. Participants discussed the importance of early planning of stakeholder involvement to ensure that all relevant parties have a seat at the table. However, the more diverse the participation, the more diverse the issues and the more extensive the meeting requirements will be. Participants emphasized that meeting costs is not a trivial issue and can be disruptive, even if beneficial in the long run.

Information Research

Participants also noted that staff time may also be required to investigate relevant issues, and to collect relevant information. Participants recognized that many of these research activities are needed by utilities and would occur regardless of the level of involvement in the TMDL. Workshop participants discussed four groups of information that may be needed: information on source water quality; contaminant source identification and control; regulations; and TMDL impacts on the utility.

TMDLs inherently require comprehensive receiving water quality data to develop a reliable TMDL. However, the analysis of these data is done in the context of developing a TMDL and typically not in the context of drinking water issues. Utilities will often want to re-analyze the water quality data and search out other sources of additional data relevant to their specific utility issues.

Another area of strong interest to utilities is identification and quantification of pollutant sources and their control. This information research has many elements common to the Source Water Assessment Program (SWAP) but more is needed for a TMDL. Research on sources may include not only compiling an inventory of potential sources (e.g., CSOs, SSOs, agricultural sources), but also researching farming practices in the watershed and the relationship to finished water quality, as well as the types of actions and controls that may be implemented for a variety of pollutant sources. The reliability and effectiveness of various BMPs is invariably a topic of TMDL debate and can be an important issue in ensuring a protected water supply. These investigations can be time consuming but may produce quite valuable information.

An understanding of relevant regulations may also require legal and administrative research including investigations on regulations that may impact the utility, a review of relevant water quality standards, and in some cases a review of the need for site specific standards. In particular, participants discussed that utilities will want to explore regulatory vehicles to control contaminant sources and explore how the TMDL program can be employed to address contaminants of concern not directly addressed by the TMDL. Utilities need to also understand roles and responsibilities of all the various regulatory stakeholders.

Another costly but high value need discussed was the investigative effort needed to analyze the implication of an implemented TMDL. This could mean added regulation for the utility or could beneficially result in cost savings or increases resulting from changed source water quality and the related treatment requirements. In both cases, utilities will need to fully understand the cost and benefit implications of the TMDL in order to strategize their investment in the process.

Data Collection and Analysis

Participants also discussed the potential cost (and benefit) of utilities directly participating in data collection and analysis. Data required to conduct a scientifically sound TMDL may often be quite limited. Calculating a reliable TMDL can require comprehensive data not often available. Utilities, if involved in the process, may decide to contribute data collection and analysis in order to enhance the quality and reliability of the TMDL to better meet their specific needs. Monitoring activities to support the TMDL may include sampling, laboratory analysis, data management, and data analysis. Monitoring data that may be needed include instream water quality data, hydrology data, and data on loadings including nonpoint sources. There may also be a need to collect land use data and develop GIS coverage for the watershed. Although certainly not the responsibility of the utility, the expense can be warranted when considering the results of a poorly developed TMDL that does not adequately protect and restore water quality or improperly regulates the utility.

For some TMDLs, there will also be a need for other specialized studies such as comprehensive planning studies or technical studies such as computer modeling analyses. Special field studies may include algal studies or septic tracer studies. Again, utilities may be faced with the option of a TMDL with limited scientific support, or conducting necessary studies at their own initiative and cost.

Risk of Outcomes

Workshop participants also discussed how the outcomes of a TMDL may be uncertain, posing significant risks, and how the process may open a "Pandora's Box." These outcomes may be a direct consequence of the TMDL regardless of the role played by the utility in the process, or they may be a consequence of the higher visibility of the utility through participation. One possible outcome is that the TMDL does not solve the problem, or there may be political damage or other negative fallout from the TMDL process. Other potential outcomes may include controls that are costly or restrictive to the utility, and higher public scrutiny of utility operations. One potential outcome of the TMDL process is that it may not solve the problem, after the utility has invested significant funds in the process. For example, there may still be a bacteria, or metals or pesticide problem to deal with after the TMDL has been completed. This may occur because of uncertainty in the analysis, lack of implementation, or other unexpected outcomes. Alternatively, the TMDL may identify a previously unknown threat, that is not currently treatable with available technologies, or is costly to treat. These types of outcomes may also increase public concern.

The TMDL may result in controls that are costly or restrictive to utilities, such as treatment of discharges, or restriction of source water withdrawals. For example, one possible outcome of the dissolved oxygen TMDL scenario is that the utility will be required to identify a new source, or reduce their withdrawal rate, to mitigate their effect on the baseflow and consequent impact on dissolved oxygen levels in the summer. However, as discussed earlier, a benefit of involvement in the TMDL process is that it provides an opportunity to have a voice, and ensure that any controls are fair and feasible.

Participation in the TMDL may result in higher public scrutiny of the utility, although this may occur regardless of the level of involvement of the utility. There may be an increase in public apprehension about the safety of the water supply as a result. The process may also highlight other water quality issues that are not directly related to the TMDL, such as pesticides. This may put greater pressure on the utility for additional treatment or finding alternate sources, rather than keeping the focus on the TMDL and source water assessment and protection. Public relations may also suffer if the costs of TMDL involvement mentioned above result in higher costs for ratepayers.

Finally, there may be political damage in the aftermath of the TMDL, or other unanticipated outcomes. For example, a utility in a farming community may advocate for agricultural controls to reduce sediment and nutrient loads in source water, and this may conflict with a desire by others to maintain a low profile.

SUMMARY

In summary, the workshop participants recognized the importance of drinking water utility participation in the TMDL process, and they also identified potential costs and risks. Participants were also asked for example case studies that highlight these benefits and costs, and these are listed in Appendix F. During the session on future research needs, summarized in the following chapter, participants recommended investigations of selected case studies as part of future research investigations. An in-depth examination of these case studies is outside the scope of this effort; rather these case studies are listed for reference in any future investigations.

CHAPTER 5
INFORMATION AND RESEARCH NEEDS

The final workshop session was directed at identifying information and research needs specific to drinking water utility involvement in the TMDL program. In a brief presentation (see Appendix H), participants were provided with some background information to set the context and provide background for discussion. Participants were then asked, in an open discussion format, for their ideas on priority needs specific to drinking water utilities and the TMDL program.

In the overview before the discussion, highlights of some recent national activities directed at improving the TMDL program were provided. These activities included:

- The 1996 Federal Advisory Committee Act (FACA) to evaluate US EPA and the States' implementation of the TMDL program, and recommend improvements (US EPA 1998)
- A National Academy of Sciences review of the adequacy of methods and approaches of the TMDL program (NRC/NAS 2001)
- US EPA's 2002 report on research needs to improve the TMDL Process (US EPA 2002c)
- Recent Water Environment Research Foundation (WERF) research directed at improvements to the TMDL program (Freedman et al. 2003a; Thornton and Freedman 2003; Freedman et al. 2003b)
- A recent review of research needs related to the TMDL program published in Water Environment & Technology (Freedman, Dilks, and Larson. 2003c)

While many of the needs identified through these efforts are relevant to the drinking water industry, these national efforts were directed at improvements to the TMDL program as a whole. Workshop participants were asked to focus specifically on needs of highest concern to drinking water utilities. As a starting point for discussion, needs that were identified in pre-workshop surveys and during the preceding workshop sessions were compiled and presented. These identified needs were grouped into the following six categories, and are listed in the presentation in Appendix H:

- TMDL analysis
- Standards and the TMDL process
- Implementation of controls
- Program integration
- Watershed management
- Educating drinking water utilities

Participants reviewed this list, and then went on to discuss priority needs. The results of these discussions are presented below. It is important to note that discussions were more of a brainstorming exercise than a strategic planning session. No ranking or sorting of needs was performed. Hence this outcome of discussions is more of a summary of needs discussed and not a prioritization of highest needs. Also it should be noted that some of the needs represented clear research whereas others are more of a compilation of information and an educational outreach process. Both aspects were deemed important.

TMDL ANALYSIS

The general topic areas of research described by the participants related to improving the quality and care of perfecting the TMDL calculations. Issues identified related to watershed characteristics, data collection, pollutant source identification, and modeling guidance. Most of these issues were not only relevant to drinking water utilities but also other stakeholders.

Watershed Characterization of Source Waters

Improvements in watershed characterization were identified as a need by workshop participants. This includes better identification of sources through better land use characterization but most importantly, they emphasized the need to better identify locations and sources of water intake in the TMDL impaired waters. Information and GIS technology is readily available to geographically relate impaired watersheds needing TMDLs and vulnerable drinking water sources, but this generally has not been done. Participants saw an urgent need to compile this information. This process would also naturally lead to the need to explore issues of watershed scale when translating 303(d) impaired waters into a TMDL watershed assessment.

Improving Data Collection and Sharing

Workshop participants also saw a need to improve the collection of water quality data in impaired waters, a need highlighted in the other TMDL studies discussed earlier. In particular the participants recognized a need to integrate various disjointed monitoring programs to better serve TMDL needs and protect drinking water sources. This involved not only an inventory of the sources of data but also a framework to share data and develop integrated monitoring programs. Specifically the participants called for an inventory of data sources and matching of these sources to drinking water concerns. The participants also spoke of the need for better guidance on survey design, especially wet weather surveys in order to better measure episodic sources that could be a risk to drinking water utilities. In addition, they spoke of a need to advance the use of continuous remote monitoring devices that could provide better data for TMDL analysis and a better ability to continuously monitor threats to drinking water supplies.

Source Identification

Monitoring for source identification was also an important concern to workshop participants. In particular, they spoke often of the need for better source identification for pathogens like *E. coli*, *Giardia* and *Cryptosporidium*. This includes advancing analytical techniques like "fingerprinting" technologies, or surrogate measures for identifying sources, to developing better sampling program designs to capture and identify ongoing as well as episodic sources.

Improve Modeling

Aside from data collection, the participants spoke of needs to improve access, use and available technology for models used in TMDL development. This too was highlighted in the other prominent TMDL program assessments described earlier. However, the participants' interest was different than other groups. The participants did speak of the need for data calibration/verification guidelines, an issue identified by others. However, the workshop participants spoke uniquely about the need to develop a "buyer's guide," so to speak for modeling. A consumer-oriented description of models, capabilities, and limitations that would guide the reader through a process of how to evaluate whether an agency, consultant, or researcher is providing quality work. Not a "how to guide" and not a theoretical discourse, but rather basic model information that a "consumer" of model results needs to know and a list of questions that should be asked to insure that models are properly applied to the TMDL and drinking water problems at hand. This should include ample cross references about how specific models have been used for similar problems.

Other modeling issues identified by the participants included more research on:

- The fate & transport of new pollutants of concern to drinking water industry, including pathogens, disinfection byproducts precursors, pesticides, and endocrine disrupters, and pharmaceuticals as examples;
- Improved understanding of model reliability and uncertainty; and
- Better characterization of groundwater/surface water interactions.

STANDARDS AND THE TMDL PROCESS

Standards are at the heart of the TMDL process. They are the measure by which impaired waters are identified and the TMDL mandated, and they are the target that completed TMDLs are required to meet. However, workshop participants universally agreed that the Clean Water Act standards program is not very effective in meeting the needs of drinking water utilities. They spoke of the need to improve employment of use designations, and narrative standards, as well as develop new criteria to address contamination not typically addressed in the TMDL program.

Designated Water Supply Use

At the start participants agreed that States were often ineffective in explicitly designating water bodies as drinking water sources, as well as associating that use with specific water quality criteria. Research is needed to document and publicize these shortcomings. The CWA is effective in addressing "swimmable and fishable" uses for our waters, but protecting safe drinking water, although a bonafide use, is not adequately addressed. Waters need to be designated "protected for drinking water" or alternative descriptions need to be assigned (e.g. special protected or sensitive waters) that serve to protect this use.

On the topic of designated uses, the workshop participants discussed a recent memorandum from the assistant administrator for the US EPA Office of Water (Mehan 2003) that calls for steps to ensure that waters that are used as a drinking water source have water quality standards with public water supply as a designated use, or they are protected by water quality criteria that are fully protective of that use designation. Participants recommended that a response supporting these efforts be prepared as soon as possible.

Drinking Water Standards for Source Water

Not withstanding this shortcoming, participants did not feel the CWA water quality criteria were adequately comprehensive and specific to address drinking water needs. For many drinking water concerns such as pesticides, DBP precursors, *Cryptosporidium*, *Giardia*, nitrates, and numerous trace organics for which MCLs exist, there are no corresponding surface water CWA criteria. For other parameters there are, at times, wildlife criteria (e.g., mercury and dichlorodiphenyltrichloroethane (DDT)) that are very protective of drinking waters but these are not universally applied. Research is needed to bridge the gap and develop appropriate criteria and establish methodologies to develop reliable measurement techniques and TMDL analysis for parameters of concern to drinking water utilities.

Improve Use of Narrative Standards

For many parameters of concern the CWA uses a qualitative objective called *narrative criteria* as a standard tool. These narrative standards generally speak to prohibition of "toxic substances in toxic amounts" or contaminants causing objectionable conditions such as taste and odor or nuisance algal conditions. However, the supporting guidance to translate these objections to specific numeric criteria in a TMDL is inadequate, especially for nutrients, and research is needed to address this need. For nutrients, research has been conducted but the focus is on ecology, not drinking water protection. The same is the case for toxic substances. Research is needed to relate CWA narrative standards to issues important to both ecology and drinking water protection.

Period Averaging for Standards

Another issue identified was that many CWA standards generally are period averages. Bacteria standards are often expressed as a geometric mean, and other parameters may be assessed in a daily, weekly or monthly context. In contrast drinking water must be protected 24/7/365. This disconnect needs to be recognized and addressed. Standards for surface waters must protect both aquatic life and public health with appropriate criteria.

State Pilot Demonstrations

The participants felt strongly that leadership on many of the above issues will first have to come from the states. The workshop participants were very enthused about the possibility of identifying a specific state to explore these standards and issues. They viewed such a pilot as a high priority. Arizona and Kansas were suggested as two possible pilot states.

IMPLEMENTATION OF CONTROLS

Ultimately TMDLs are only beneficial if controls are implemented and water quality improved. On this topic, workshop participants identified a variety of research needs, not all specific to drinking water. These include information needs related to the effectiveness of controls, tools for encouraging or regulating controls, promoting non-traditional management approaches and controlling sources not regulated by the Clean Water Act.

Controlling Nonpoint Sources

A common concern among water quality professionals is the inability of the TMDL program to realize control of nonpoint source as well as the uncertainty of proposed best management practices to achieve desired results. Workshop participants identified a need for research in both areas. First they saw a need to explore regulatory tools and incentive programs such as trading to promote control of nonpoint sources not under regulation. Particular emphasis was focused on inadequate controls for agriculture sources. However, participants were equally concerned that BMPs used for nonpoint sources would not achieve desired results. In particular, they felt better research was needed on the effectiveness and cost-benefits analysis of BMPs to reduce pollutants of concern specifically to drinking water issues. Effectiveness of most BMPs is highly uncertain, but when good data are available it is focused on sediment issues, not other issues particular to drinking water such as pesticide reduction.

Unregulated Sources and Stressors

The participants also discussed another gap in the Clean Water Act TMDL program, which did not regulate all contributors to impairment such as atmosphere sources and flow regulation/management. Participants felt research was needed to identify the importance of these factors and methods for achieving their control or better management.

PROGRAM INTEGRATION

A universal complaint of the workshop participants was that the CWA and TMDL requirements are not integrated explicitly or implicitly with other regulatory programs. This includes not only the Safe Drinking Water Act but other regulatory programs affecting water including the Clean Air Act, various programs affecting farm practices and use of pesticides and herbicides, and programs affecting Federal land management as examples.

Road Map to Program Integration

The workshop participants felt a need to initially develop a "guide" that examines the interrelationships between programs. This would require not only a summary of programs affecting water supplies and water quality, but also cross definitions of terminologies, nature of regulatory and incentive programs, and most importantly the "players", the points of contact at Federal and state agencies that control these programs. The participants felt that through the use of this kind of "road map," TMDL participants could be effective through individual efforts in integrating programs not integrated by legislation or regulation. Participants also felt that there is definitely a need to educate regulators involved in CWA programs about drinking water utility operations and concerns.

Integrating Source Water Assessment and Protection Programs and TMDL

The participants also often discussed in the workshop the many similarities between the CWA TMDL program and the SDWA source water assessment program (and any follow up Source Water Protection Plan (SWPP). Participants felt that many synergies could be accomplished by integrating these programs. A research study or pilot study exploring this would be valuable, or documentation of case studies where these synergies have been achieved.

WATERSHED MANAGEMENT

The TMDL program is in essence a regulatory adaptation of watershed management, a concept strongly supported by all the workshop participants. However, there was also general agreement that to protect drinking water supplies, watershed management needs to expand beyond traditional TMDL analysis. The topics highlighted were use of adaptive management, application of anti-degradation and the need for effective management of Federal lands.

Adaptive Management

On the issue of adaptive management, the participants felt that TMDL analyses are often constrained by large uncertainties in standards, quantification of pollutant loads, effectiveness of controls, and ability to precisely relate sources to a quantitative objective. Participants felt research was needed to promote the use of adaptive management in the TMDL process. We need to view TMDLs not as a single number determined once, but rather a process of continual improvement as better information is obtained. This is especially important for drinking water utilities where the CWA objectives are not focused, the relevant controls have uncertain effectiveness, and especially where pollutants of concern extend beyond traditional TMDL analysis.

Anti-degradation

Of special interest to participants was the innovative idea of applying the CWA anti-degradation policies to limit further degradation in a watershed and promote better land uses such as low impact development. The workshop participants felt research on how to take advantage of the CWA anti-degradation policies both in the context of TMDL and otherwise might be an avenue to protect drinking water sources beyond activities already occurring in the CWA. Anti-degradation and adaptive management can be very complementary approaches.

Historically, anti-degradation has been used mostly to limit increases in point sources for pollutants regulated as effluent limits. The workshop participants discussed how the policy could be employed to regulate other emerging pollutants and control future watershed development. However, case studies and research are needed to explore this important approach.

Managing Federal Lands

Last in this area, participants noted the enormous tracts of land governed by Federal Agencies, including US Forest Service, Bureau of Land Management, and Department of Defense as examples. Many of these areas include tributaries to important water supplies. The participants outlined a need to explore mechanisms to explicitly control land practices in the Federal lands to meet TMDL objectives and protect drinking water sources. Currently these lands are largely outside the regulatory reach of the TMDL program.

EDUCATING DRINKING WATER UTILITIES

Although most of the workshop participants were particularly well acquainted with the TMDL program (many were chosen for the workshop just for this reason), there was a strong feeling that most other utilities (especially small utilities) would not be so familiar. Therefore, there was generally agreement that strong efforts need to be focused on getting utilities both acquainted with the TMDL program and encouraged to become involved. Besides publishing the findings of the workshop, the participants suggested the preparation of a TMDL guidance manual for utilities with a special section for small utilities, and case studies and outreach through the development of related publications, workshops, and conferences.

TMDL Manual for Utilities

The workshop participants felt there was a need to prepare a simple guidance manual that would help utilities determine the need to be involved in a TMDL and how to be effectively involved. The primer would be written for utilities from their perspective, outlining the terminology, players, and steps involved in a TDML. This would include a series of questions that would guide a utility through the evaluation of deciding to become involved in a TMDL and the type of consideration and information needed. This involves defining the key questions utilities need to ask to insure that their interests are best protected. Distribution of the primer would include posting it on a web page to reach a broad audience.

Such a TMDL primer might include example case studies that describe situations where drinking water utilities have been involved in TMDLs or watershed management activities to protect and improve source water. Workshop participants felt that investigations of case studies would be valuable in better understanding the issues, benefits, and costs associated with participation in TMDLs. During all sessions of the workshop, participants were asked for case studies that highlight the issues or topics under discussion. A list of the case studies suggested by participants is provided in Appendix F. An in-depth examination of these case studies is outside the scope of this effort; rather these case studies are listed for reference in any future investigations.

Information Outreach

The second component of utility education recommended by the participants was widespread publicity of TMDL information through professional publications, conference papers, workshops, teleconferences, and web-based outreach. The participants felt strongly that the current workshop was highly beneficial to them, even recognizing their relatively high knowledge of TMDLs. Spreading the word nationally on the importance of TMDLs to drinking water utilities is paramount if utilities expect to take best advantage of the programs to protect and improve their water supplies.

One specific idea related to information outreach that participants explored was the need for a workshop framework or template that could be developed for use at various national conferences, but more importantly packaged in a format suitable to send it "on the road" for presentation at numerous regional and state sessions. Another idea was to prepare a short article for a publication such as Mainstream, Opflow, or Journal American Water Works Association (AWWA) that poses a series of questions and provides short answers related to the TMDL program. The questions and answers would be developed to provide basic information about the TMDL program, and the impacts on drinking water utilities. This was identified as a high priority activity that should be conducted as soon as possible.

RESEARCH TEAM RECOMMENDATIONS

In the context of the workshop, participants did not establish a prioritization of information and research needs, only an inventory. The list is large and diverse. Prioritizing the needs was beyond the objectives of the workshop, but would be useful. Therefore, the research team has identified what it views as the highest priority needs among those identified at the workshop. Brief descriptions of potential research projects which are not in a specific order are provided in the following paragraphs. While not all projects are directly related to TMDLs, they all relate to the development of a tool set for utilities to protect source waters, drawing from CWA programs. Each research need listed is justified by the context of the problem, recommended study, and benefit to the drinking water utilities.

1. Anti-degradation Policies

Context of problem: Water quality standards require a designated use, criteria that meet the uses, and an anti-degradation policy. There is a need for research on how state anti-degradation policies across the U.S. are applied and how they can be applied to protect source waters. TMDL focuses on restoration for listed impairments whereas, anti-degradation could focus on preventive protection.

Recommended study: Project would review state policies and evaluate their applications for drinking water.

Benefit to drinking water utilities: A possible CWA tool to protect the viability of water supplies where they are vulnerable to degradation.

2. Program Integration

Context of problem: The CWA and SDWA as well as other Federal regulation programs such as the Toxic Substances Control Act (TSCA) and federal and state land management programs are not currently integrated. There is an expressed need for a comprehensive watershed program that goes beyond the TMDL program and includes the designation of all drinking water sources. This investigation would identify the tools that are available to protect source waters. The project could go beyond just CWA and SDWA to include many other regulatory and non-regulatory Federal programs affecting water quality.

Recommended study: The project would survey and inventory relevant programs. A more detailed approach might involve a case study of an individual state and how they are integrating programs.

Benefit to drinking water utilities: The project will move toward an integrated program to protect source waters.

3. Water Quality and Drinking Water Standards

Context of problem: Differences in standards under the CWA and SDWA include parameters, frequency and duration and uses. For example, MCLs may exist with no parallel standards in the CWA and CWA standards are often very different than SDWA standards. Further, the relationship between 303(d) impaired waters and designated uses for drinking water is currently unknown.

Recommended study: This project would survey states to determine how they are addressing this disconnect. Research would focus on identifying all the different regulations, and provide a discussion of how all of the programs could be coordinated.

Benefit to drinking water utilities: Identification of which existing water quality standards utilities can use to protect sources, and which additional parameters could be included. It leads towards a comprehensive resource management.

4. Narrative Standards

Context of problem: Narrative standards such as *"no toxic substances in toxic amounts"* are a potential tool that can be used as a lever to get additional source waters protected.

Recommended study: Investigation would look at how narrative standards can be most effectively applied, to protect drinking water sources where specific CWA standards are not violated or non-existent for the for the parameters of concern.

Benefit to drinking water utilities: Utilities will be better informed about how they might use narrative standards to protect and restore sources.

5. Data Needs for TMDLs

Context of problem: Drinking water utilities currently lack information on the types of data and information needed for TMDLs. A "how to" manual would be too generic, as surveys are very site specific. Instead, utilities would benefit from an inventory of needs.

Recommended study: This project would identify the data and information required for TMDLs, and cross link the needs to information on where the data can be obtained.

Benefit to drinking water utilities: Utilities currently lack this type of information and the product would provide a reference that would enable informed involvement in TMDLs.

6. Pathogen Fingerprinting

Context of problem: Fingerprinting identifies the sources of pathogens so utilities can trace contamination in their source waters. This project would provide an overview of the state of the science including current laboratory procedures.

Recommended study: Conduct surveys focused on current state of the science.

Benefit to drinking water utilities: Current information on the status of this tool and its use in protecting source waters.

7. TMDL "Buyer's Guide"

Context of problem: Drinking water utilities getting involved in TMDLs need to understand the types of questions to ask to ensure successful outcomes. The outcome of this project would be the most important questions with discussion for drinking water utilities to ask when they are involved in a TMDL.

Recommended study: Develop and address the essential questions when utilities become involved in a TMDL.

Benefit to drinking water utilities: Project will ensure that TMDLs are done well and that utilities understand their options.

8. Modeling Fate and Transport of Emerging Pollutants

Context of problem: There are many emerging contaminants of concern that we don't currently have much experience with. Utilities need information on their vulnerability to upstream sources.

Recommended study: This project would involve a literature review and screening level calculations to identify sources of emerging pollutants and the document their behavior in the environment.

Benefit to drinking water utilities: Assessment of vulnerability to upstream sources, specifically with respect to emerging pollutants.

9. Adaptive Management/Model Uncertainty

Context of problem: For most TMDLs, there is not enough data and high uncertainty in model results and outcomes. An adaptive management approach allows steps to be taken and adjusted as knowledge is gained.

Recommended study: The project will explore these issues, examine case studies where an adaptive management approach has been used, and identify formulas for success.

Benefit to drinking water utilities: The project will assist utilities in making sure that adaptive management meets expectations and results in protection of source waters, and is not used as a means of avoiding controls.

CHAPTER 6
SUMMARY AND RELEVANCE TO THE DRINKING WATER COMMUNITY

Currently, SDWA and CWA regulatory and management efforts to protect the quality of surface waters are disjointed. In particular, SWAP plans are being developed by states to help identify source of contamination to drinking water supplies. In contrast, the TMDL and NPDES regulatory programs which are designed to quantify and control pollution sources to surface waters are being conducted, generally, unrelated to drinking water protection. Conceptually it seems that all stakeholders would benefit from the cooperation and integration of source water protection efforts like SWAP with CWA regulation efforts to reach designated uses through the TMDL requirements. Water utilities have historically focused on SDWA but they need to understand the essence and detail of the TMDL program and identify opportunities for direct involvement and improvement. Drinking water utility involvement with TMDLs can be both as a protector of the source water or as a possible contributor to non attainment of designated uses. Therefore, drinking water utilities are in a position that the two sets of regulations can interact and at times be at odds.

The results of this research project have numerous practical benefits to the drinking water community. This study presents important information about the TMDL regulations and the relationships of TMDLs to SDWA regulations. The material developed can help guide drinking water utilities into both the liabilities presented by the TMDL requirements and the way TMDLs can be used as part of a source water protection program. This research also identified benefits and risks to water utilities from the establishment of TMDLs and future TMDL research needs that will support the drinking water utilities In addition, this project has documented several issues and barriers that the drinking water utilities need to consider in order to be effective stakeholders in the TMDL process. These issues were consolidated into six major areas representing data collection and analysis, CWA and SDWA water quality standards, implementation of controls, interagency relationships, public outreach, and watershed comprehensive planning.

Specifically, utilities that are facing the possibility of involvement in a TMDL process should consider the following points in preparation for their involvement:

1. Is the utility likely to be a stakeholder responsible for non attainment or using the TMDL to protect the water body?
2. Who are the other likely stakeholders? What roles are they likely to represent? What other utilities are involved? What other agencies are likely to be interested, involved or in authority positions?
3. What other stakeholders could supply resources to support the TMDL process: money, data, technical expertise, monitoring, and management?
4. Who are the responsible government agencies for conducting and enforcing the TMDLs?
5. What are the designated uses for the water body?
6. What are the CWA, SDWA or state standards that are likely to be in play?
7. Is the water body considered impaired? If so for what designated uses?

8. What are the sources of impairments?
9. On what basis was the water body classified as impaired?
10. What data are available for the water body and for the surrounding area?

    - Are the data available?
    - Are the data in electronic format?
    - Is data documentation available describing sample design, purpose, methodology, etc?

11. What models are available for the water body (specifically or in general)?
12. What on-going monitoring is there on the water body now?
13. What other water bodies influence the water quality?
14. Which scenarios, actions, or management plans can be applied for reaching compliance with designated uses?

A qualitative analysis evaluated and assessed benefits and costs for water utilities that become involved in the TMDL process. In summary, the outcomes of this analysis strongly endorsed the importance of drinking water utility participation in the TMDL process and recognized any negative consequences would occur with or without utility participant. The types of the benefits that were identified included reduced public health risk, reduced treatment requirement and costs, appropriate control measures, good public relations, more efficient regulatory interactions, improving water quality information, and more effective future planning. On the other hand, costs were mainly associated with participation in meetings and negotiations, and monitoring.

This project also identified future research studies that will benefit drinking water utilities. These needs will be valuable for all utilities, large and small, regardless of their prior level of sophistication and involvement in regulatory and water quality management issues. The identified needs were grouped in six major areas including TMDL analysis, water quality standards and the TMDL process, implementation of controls, program integration, watershed management, and educating drinking water utilities. Some of the recommended future studies included assessment of state and federal program integration; interface of CWA and SDWA water quality standards; identification of TMDL data needs for drinking water utilities; compilation of information and questions to be asked on modeling, data, allocation and BMPs; and modeling fate and transport of emerging pollutants.

The TMDL requirements represent both a challenge and an opportunity for drinking water utilities. These utilities may find themselves involved in a stakeholder process requiring resources for monitoring, data analysis, interpretation and mitigation. While these encounters may require unplanned resource expenditures and issues they are also an opportunity to interact with the public and identify partners in the maintenance of the environment and water resource. On the other hand, the water utilities may find that TMDLs can be effectively used to organize the community to protect water sources and a means by which the utility can get other entities to take responsibility for monitoring and mitigation. TMDLs can also be a tool for a drinking water utility to get responsible parties to discontinue practices that are harmful to drinking water resources.

REFERENCES

Conservation Ontario. 2001. *Valuing Benefits from Water Management. A Supplement to the Report "The Importance of Watershed Management in Protecting Ontario's Drinking Water Supplies".* Ontario, Canada: Conservation Ontario [Online]. Available: <http://conservation-ontario.on.ca/projects/walkerton/files/CObenefits.pdf>. [cited October 4, 2004].

Dearmont, D, McCarl, B., and Tolman, D. 1998. Costs of Water Treatment due to Diminished Water Quality: A Case Study in Texas. *Water Resources Research*, 34(4): 849-853.

Ernst, C., Gullick, R., and Nixon, K. 2004. Protecting the Source: Conserving Forests to Protect Water. *Opflow*, 30(5):1-7. Denver, Colo.: AWWA.

Freedman, P.L., Dilks, D.W., Holmberg, H.P., Moskus, P., McBridge, G. Hickey, C., Smith, D. , and Stripin, P.L.. 2003a. *Navigating the TMDL Process: Method Development for Addressing Narrative Criteria*. Project No. 01-WSM-1, Alexandria, Virginia: WERF.

Freedman, P.L., Larson, W.M., Dilks, D.W., Schechter, D., Nemura, A., Naperala, T., DePinto, J.V., Prothro, M.G., Boese, G.W., Dettelbach, A., Nothman, L., Thornton, K., Ford, D., Massirer, P., Stevens, K.B., Sobrinho, J.A.H. 2003b. *Navigating the TMDL Process: Evaluation and Improvements*. Project No. 00-WSM-1, Alexandria, Virgina:WERF.

Freedman, P.L., Dilks, D. W., and Larson, W.M. 2003c. Improving TMDLs. *Water Environment & Technology*, 15(6):22-26.

Mehan, III, G. T. 2003. Memorandum: Effective Use of Water Quality Standards to Protect Sources of Drinking Water. US EPA, Office of Water, Washington, DC [Online]. Available: <http://biosystems.okstate.edu/waterquality/hot_topics/EPA-Oct1-2003.pdf>. [cited October 4, 2004].

National Research Council of the National Academy of Sciences (NRC/NAS). 2001. *Assessing the TMDL Approach to Water Quality Management*. National Academy Press, Washington, DC [Online]. Available: <(http://www.nap.edu/catalog/10146.html)>. [cited October 4, 2004].

Schechter, D., Freedman, P.L., and Dilks, D. W. 2000. Leveraging Water Quality Assessment Resources - Why TMDLs and SWAPs Should be Performed Together. In *Proc. of the 73th Water Environment Federation Technical Exhibition and Conference*. Anaheim, California: WEF.

Thornton, K.W., and Freedman, P.L. 2003. *Navigating the TMDL Process: Listing and Delisting*. Project No. 00-WSM-2, Alexandria, Virginia: WERF.

US EPA. 1991. *Guidance for Water Quality-Based Decisions: The TMDL Process.* EPA 440/4-91-001, US EPA, Office of Water, Washington, DC [Online]. Available: <http://www.epa.gov/owow/tmdl/decisions>. [cited October 4, 2004].

US EPA. 1998. *Report of the Federal Advisory Committee on the Total Maximum Daily Load (TMDL) Program.* EPA 100-R-98-006, US EPA, Office of Water, Washington, DC [Online]. Available: <http://www.epa.gov/owow/tmdl/advisory.html>. [cited October 4, 2004].

US EPA. 1999a. *Draft Guidance for Water Quality-based Decisions: The TMDL Process (Second Edition).* EPA 841-D-99-001, US. US EPA, Office of Water, Washington, DC [Online]. Available: <http://www.epa.gov/owow/tmdl/tmdlguid.pdf>. [cited October 4, 2004].

US EPA. 1999b. *Protocol for Developing Nutrient TMDLs.* First Edition, EPA 841-B-99-007. US EPA, Office of Water, Washington, DC [Online]. Available: <http://www.epa.gov/owow/tmdl/nutrient/pdf/nutrient.pdf>. [cited October 4, 2004].

US EPA. 1999c. *Protocol for Developing Sediment TMDLs.* First Edition, EPA 841-B-99-00, US EPA, Office of Water, Washington, DC [Online]. Available: <http://www.epa.gov/owow/tmdl/sediment/pdf/sediment.pdf>. [cited October 4, 2004].

US EPA. 1999d. *Protecting Sources of Drinking Water - Selected Case Studies in Watershed Management.* EPA 816-R-98-019, US EPA, Office of Water, Washington, DC [Online]. Available: <http://www.epa.gov/safewater/swp/swpcases.pdf>. [cited October 4, 2004].

US EPA. 2000. *1998 National Section 303(d) List Fact Sheet.* US EPA, Office of Wetlands, Oceans and Watersheds, Washington, DC [Online]. Available: <http://oaspub.epa.gov/waters/national_rept.control >. [cited October 4, 2004].

US EPA. 2001. *Protocol for Developing Pathogen TMDLs.* First Edition. EPA 841-R-00-002. US EPA, Office of Water, Washington, DC [Online]. Available: <http://www.epa.gov/owow/tmdl/pathogen_all.pdf>. [cited October 4, 2004].

US EPA. 2002a. *2002 Integrated Water Quality Monitoring and Assessment Report Guidance Memorandum.* US EPA, Office of Wetlands, Oceans and Watersheds, Washington, DC [Online]. Available: <http://www.epa.gov/owow/tmdl/2002wqma.html>. [cited October 4, 2004].

US EPA. 2002b. *Consolidated Assessment and Listing Methodology: Toward a Compendium of Best Practices.* US EPA, Office of Wetlands, Oceans and Watersheds, Washington, DC [Online]. Available: <http://www.epa.gov/owow/monitoring/calm.html>. [cited October 4, 2004].

US EPA. 2002c. *The Twenty Needs Report: How Research Can Improve the TMDL Program.* EPA 841-B-02-002, US EPA, Office of Water, Washington, DC [Online]. Available: <http://www.epa.gov/owow/tmdl/20needsreport_8-02.pdf>. [cited October 4, 2004].

LIST OF ABBREVIATIONS

AWWA	American Water Works Association
AwwaRF	Awwa Research Foundation
BMP	Best Management Practice
CWA	Clean Water Act
CALM	Consolidated Assessment and Listing Methodology
CSO	Combined Sewer Overflow
DBP	Disinfection by-Product
ESA	Endangered Species Act
FACA	Federal Advisory Committee Act
GIS	Geographic Information System
NAS	National Academy of Sciences
NPDES	National Pollutant Discharge Elimination System
NPS	Nonpoint source
NRC	National Research Council
MCL	Maximum Contaminant Level
MOS	Margin of Safety
NACEPT	National Advisory Council for Environmental Policy and Technology
NGO	Non-Governmental Organization
NOM	Natural Organic Matter
NYC	New York City
ORD	Office of Research and Development
PCB	Polychlorinated Biphenyl
SDWA	Safe Drinking Water Act
SSO	Sanitary Sewer Overflow
SWAP	Source Water Assessment Program
SWPP	Source Water Protection Plan
THM	Trihalomethane
TMDL	Total Maximum Daily Load
TSCA	Toxic Substances Control Act
US EPA	United States Environmental Protection Agency
UAA	Attainability Analysis
USACE	USA Corps of Engineers
WERF	Water Environment Research Foundation
WQS	Water Quality Standard

IWA Publishing's authorised EU representative for General Product Safety Regulations is Diane D'Arras, 15 rue Duret, 75116 Paris, France, e-mail: safety@iwap.co.uk.

Printed and bound by CPI Group (UK) Ltd, Croydon, CR0 4YY
17/09/2026
02218802-0009